The Science of Fingerprints, by

The Science of Fingerprints, by

Federal Bureau of Investigation John Edgar Hoover

This eBook is for the use of anyone anywhere at no cost and with almost no restrictions whatsoever. You may copy it, give it away or re-use it under the terms of the Project Gutenberg License included with this eBook or online at www.gutenberg.org

Title: The Science of Fingerprints Classification and Uses

Author: Federal Bureau of Investigation John Edgar Hoover

Release Date: August 10, 2006 [EBook #19022]

Language: English

Character set encoding: ISO-8859-1

*** START OF THIS PROJECT GUTENBERG EBOOK THE SCIENCE OF FINGERPRINTS ***

Produced by Jason Isbell, Linda Cantoni, and the Online Distributed Proofreading Team at http://www.pgdp.net

THE SCIENCE

OF

FINGERPRINTS

Classification and Uses

UNITED STATES DEPARTMENT OF JUSTICE

FEDERAL BUREAU OF INVESTIGATION

John Edgar Hoover, *Director*

INTRODUCTION

This booklet concerning the study of fingerprints has been prepared by the Federal Bureau of Investigation for the use of interested law enforcement officers and agencies, particularly those which may be contemplating the inauguration of fingerprint identification files. It is based on many years' experience in fingerprint identification work out of which has developed the largest collection of classified fingerprints in the world. Inasmuch as this publication may serve as a general reference on classification and other phases of fingerprint identification work, the systems utilized in the Identification Division of the Federal Bureau of Investigation are set forth fully. The problem of pattern interpretation, in particular, is discussed in detail.

Criminal identification by means of fingerprints is one of the most potent factors in obtaining the apprehension of fugitives who might otherwise escape arrest and continue their criminal activities indefinitely. This type of identification also makes possible an accurate determination of the number of previous arrests and convictions which, of course, results in the imposition of more equitable sentences by the judiciary, inasmuch as the individual who repeatedly violates the law finds it impossible to pose successfully as a first, or minor, offender. In addition, this system of identification enables the prosecutor to present his case in the light of the offender's previous record. It also provides the probation officers, parole board, and the Governor with definite information upon which to base their judgment in dealing with criminals in their jurisdictions.

From earliest times fingerprinting, because of its peculiar adaptability to the field, has been associated in the lay mind with criminal identification to the detriment of the other useful phases of the science. However, the Civil File of the Identification Division of the Federal Bureau of Investigation contains three times as many fingerprints as the Criminal File. These civil fingerprints are an invaluable aid in identifying amnesia victims, missing persons and unknown deceased. In the latter category the victims of major disasters may be quickly and positively identified if their fingerprints are on file, thus providing a humanitarian benefit not usually associated with fingerprint records.

The regular contributors who voluntarily submit fingerprints to the Federal Bureau of Investigation play a most important role in the drama of identification. Their action expands the size of the fingerprint files, thereby increasing the value of the files to all law enforcement agencies. Mutual cooperation and efficiency are resultant by-products.

The use of fingerprints for identification purposes is based upon distinctive ridge outlines which appear on the bulbs on the inside of the end joints of the fingers and thumbs. These ridges have definite contours and appear in several general pattern types, each with general and specific variations of the pattern, dependent on the shape and relationship of the ridges. The outlines of the ridges appear most clearly when inked impressions are taken upon paper, so that the ridges are black against a white background. This result is achieved by the ink adhering to the friction ridges. Impressions may be made with blood, dirt, grease or any other foreign matter present on the ridges, or the saline substance emitted by the glands through the ducts or pores which constitute their outlets. The background or medium may be paper, glass, porcelain, wood, cloth, wax, putty, silverware, or any smooth, nonporous material.

Of all the methods of identification, fingerprinting alone has proved to be both infallible and feasible. Its superiority over the older methods, such as branding, tattooing, distinctive clothing, photography, and body measurements (Bertillon system), has been demonstrated time after time. While many cases of mistaken identification have occurred through the use of these older systems, to date the fingerprints of no two individuals have been found to be identical.

The background and history of the science of fingerprints constitute an eloquent drama of human lives, of good and of evil. Nothing, I think, has played a part more exciting than that enacted by the fascinating loops, whorls, and arches etched on the fingers of a human being.

[Signature: J. Edgar Hoover]

J. EDGAR HOOVER, *Director.*

CONTENTS

Chapter Page

I. The Identification Division of the FBI 1

II. Types of Patterns and Their Interpretation 5

III. Questionable Patterns 71

IV. The Classification Formula and Extensions 87

V. Classification of Scarred Patterns--Amputation--Missing at Birth 98

VI. Filing Sequence 103

VII. Searching and Referencing 109

VIII. How To Take Inked Fingerprints 114

IX. Problems in the Taking of Inked Fingerprints 118

X. Problems and Practices in Fingerprinting the Dead 131

XI. Establishment of a Local Fingerprint Identification Bureau 160

XII. Latent Impressions 173

XIII. Powdering and Lifting Latent Impressions 175

XIV. Chemical Development of Latent Impressions 177

XV. The Use of the Fingerprint Camera 184

XVI. Preparation of Fingerprint Charts for Court Testimony 190

XVII. Unidentified Latent Fingerprint File 194

CHAPTER I

The Identification Division of the FBI

The FBI Identification Division was established in 1924 when the records of the National Bureau of Criminal Investigation and the Leavenworth Penitentiary Bureau were consolidated in Washington, D.C. The original collection of only 810,000 fingerprint cards has expanded into many millions. The establishment of the FBI Identification Division resulted from the fact that police officials of the Nation saw the need for a centralized pooling of all fingerprint cards and all arrest records.

The Federal Bureau of Investigation offers identification service free of charge for official use to all law enforcement agencies in this country and to foreign law enforcement agencies which cooperate in the International Exchange of Identification Data. Through this centralization of records it is now possible for an officer to have available a positive source of information relative to the past activities of an individual in his custody. It is the Bureau's present policy to give preferred attention to all arrest fingerprint cards since it is realized that speed is essential in this service.

In order that the FBI Identification Division can provide maximum service to all law enforcement agencies, it is essential that standard fingerprint cards and other forms furnished by the FBI be utilized. Fingerprints must be clear and distinct and complete name and descriptive data required on the form should be furnished in all instances. Fingerprints should be submitted promptly since delay might result in release of a fugitive prior to notification to the law enforcement agency seeking his apprehension.

When it is known to a law enforcement agency that a subject under arrest is an employee of the U.S. Government or a member of the Armed Forces, a notation should be placed in the space for "occupation" on the front of the fingerprint card. Data such as location of agency or military post of assignment may be added beside the space reserved for the photograph on the reverse side of the card.

Many instances have been observed where an individual is fingerprinted by more than one law enforcement agency for the same arrest. This duplicate submission of fingerprints can be eliminated by placing a notation on the first set of fingerprints sent to the FBI requesting copies of the record for other interested law enforcement agencies, thereby eliminating submission of fingerprints by the latter agencies.

If a photograph is available at the time fingerprints are submitted to the FBI Identification Division, it should be identified on the reverse side with the individual's complete name, name of the department submitting, the department's number, and it should be securely pasted in the space provided on the fingerprint card. If a photograph is to be submitted at a later date, it should be held until the identification record or "no record" reply from the FBI is received in order that FBI number or fingerprint classification can be added to the reverse side of the photograph for assistance of the Identification Division in relating it to the proper record.

The FBI number, if known, and any request for special handling, such as collect wire or telephone reply, should be indicated on the fingerprint card in the appropriate space. Such notations eliminate the need for an accompanying letter of instructions.

As indicated, the FBI's service is given without cost to regularly constituted law enforcement agencies and officers. Supplies of fingerprint cards and self-addressed, franked envelopes will be forwarded upon the request of any law enforcement officer. The following types of cards and forms are available: Criminal (Form FD-249), used for both arrest and institution records; Applicant (Form FD-258); Personal Identification (Form FD-353); Death Sheet (Form R-88); Disposition Sheet (Form R-84); Wanted Notice (Form 1-12); Record of Additional Arrest (Form 1-1). An order form for identification supplies appears each month with the insert to the FBI Law Enforcement Bulletin.

In addition to its criminal identification activities, the Bureau's Identification Division maintains several auxiliary services. Not the least of these is the system whereby fugitives are identified through the comparison of fingerprints which are received currently. When a law enforcement officer desires the apprehension of a fugitive and the fingerprints of that individual are available, it is necessary only that he inform the Bureau of this fact so a wanted notice may be placed in the fugitive's record. This insures immediate notification when the fugitive's fingerprints are next received.

The fugitive service is amplified by the Bureau's action in transmitting a monthly bulletin to all law enforcement agencies which forward fingerprints for its files. In this bulletin are listed the names, descriptions, and fingerprint classifications of persons wanted for offenses of a more serious character. This information facilitates prompt identifications of individuals arrested for any offense or otherwise located by those receiving the bulletin.

Missing-persons notices are posted in the Identification files so that any incoming record on the missing person will be noted. Notices are posted both by fingerprint card and by name, or by name alone if fingerprints are not available. The full name, date, and place of birth, complete description and photograph of a missing person should be forwarded, along with fingerprints, if available. Upon receipt of pertinent information, the contributing agency is advised immediately. A section on missing persons is carried as an insert in the Law Enforcement Bulletin.

The FBI Identification Division has arranged with the identification bureaus of many foreign countries to exchange criminal identifying data in cases of mutual interest. Fingerprints and arrest records of persons arrested in this country are routed to the appropriate foreign bureaus in cases when the interested agency in the United States has reason to believe an individual in custody may have a record in or be wanted by the other nation. Similarly, fingerprints are referred to the Federal Bureau of Investigation by foreign bureaus when it seems a record may be disclosed by a search of the Bureau's records. Numerous identifications, including a number of fugitives, have been effected in this manner, and it is believed that the complete development of this project will provide more effective law enforcement throughout the world. When the facts indicate an individual may have a record in another country, and the contributor submits an extra set of his fingerprints, they are transmitted by this Bureau to the proper authorities.

In very rare cases persons without hands are arrested. A file on footprints is maintained in the Identification Division on such individuals.

In view of the fact that many individuals in the underworld are known only by their nicknames, the Identification Division has for years maintained a card-index file containing in alphabetical order the nicknames appearing on fingerprint cards. When requesting a search of the nickname file, it is desired that all possible descriptive data be furnished.

The Latent Fingerprint Section handles latent print work. Articles of evidence submitted by law enforcement agencies are processed for the development of latent impressions in the Latent Fingerprint Section. In addition, photographs, negatives, and lifts of latents are scrutinized for prints of value for identification purposes. Photographs of the prints of value are always prepared for the FBI's files and are available for comparisons for an indefinite period. Should the law enforcement agency desire additional comparisons it needs only advise the FBI Identification Division, attention Latent Fingerprint Section, and either name or submit the prints of the new suspect. It is not necessary to resubmit the evidence. When necessary, a fingerprint expert will testify in local court as to his findings. Should a department have any special problems involving the development or preservation of fingerprints at a crime scene, the experts are available for suggestions. In connection with the Latent Fingerprint Section there is maintained a general appearance file of many confidence game operators. Searches in this file will be made upon request. In furnishing data on a suspect, the agency should make sure that complete descriptive data is sent in. Photographs and other material on individuals who may be identical with those being sought will be furnished to the interested departments.

During the years many persons have voluntarily submitted their fingerprints to the Identification Division for possible use in the case of an emergency. These cards are not filed with the criminal fingerprints but are maintained separately. Such prints should be taken on the standard fingerprint form entitled "Personal Identification" (Form FD-353). No answer is given to Personal Identification fingerprint cards.

The fingerprint records of the FBI Identification Division are used liberally not only by police agencies to obtain previous fingerprint histories and to ascertain whether persons arrested are wanted elsewhere, but by prosecutors to whom the information from the Bureau's files may prove to be valuable in connection with the prosecution of a case. These records are likewise of frequent value to the judge for his consideration in connection with the imposition of sentence. Obviously, the ends of justice may be served most equitably when the past fingerprint record of the person on trial can be made known to the court, or information may be furnished to the effect that the defendant is of hitherto unblemished reputation.

It should be emphasized that FBI identification records are for the *OFFICIAL* use of law enforcement and governmental agencies and misuse of such records by disseminating them to unauthorized persons may result in cancellation of FBI identification services.

CHAPTER II

Types of Patterns and Their Interpretation

Types of patterns

Fingerprints may be resolved into three large general groups of patterns, each group bearing the same general characteristics or family resemblance. The patterns may be further divided into sub-groups by means of the smaller differences existing between the patterns in the same general group. These divisions are as follows:

I. ARCH

a. Plain arch. *b.* Tented arch.

II. LOOP

a. Radial loop. *b.* Ulnar loop.

III. WHORL

a. Plain whorl. *b.* Central pocket loop. *c.* Double loop. *d.* Accidental whorl.

Illustrations 1 to 10 are examples of the various types of fingerprint patterns.

[Illustration: 1. Plain arch.]

[Illustration: 2. Tented arch.]

[Illustration: 3. Tented arch.]

[Illustration: 4. Loop.]

[Illustration: 5. Loop.]

[Illustration: 6. Central pocket loop.]

CHAPTER II

[Illustration: 7. Plain whorl.]

[Illustration: 8. Double loop.]

[Illustration: 9. Double loop.]

[Illustration: 10. Accidental.]

Interpretation

Before pattern definition can be understood, it is necessary to understand the meaning of a few technical terms used in fingerprint work.

The *pattern area* is the only part of the finger impression with which we are concerned in regard to interpretation and classification. It is present in all patterns, of course, but in many arches and tented arches it is impossible to define. This is not important, however, as the only patterns in which we need to define the pattern area for classification purposes are loops and whorls. In these two pattern types the pattern area may be defined as follows:

The pattern area is that part of a loop or whorl in which appear the cores, deltas, and ridges with which we are concerned in classifying.

The pattern areas of loops and whorls are enclosed by type lines.

Type lines may be defined as the two innermost ridges which start parallel, diverge, and surround or tend to surround the pattern area.

Figure 11 is a typical loop. Lines A and B, which have been emphasized in this sketch, are the type lines, starting parallel, diverging at the line C and surrounding the pattern area, which is emphasized in figure 12 by eliminating all the ridges within the pattern area.

Figures 72 through 101 should be studied for the location of type lines.

[Illustration: 11]

CHAPTER II

[Illustration: 12]

[Illustration: 13]

[Illustration: 14]

[Illustration: 15]

[Illustration: 16]

[Illustration: 17]

[Illustration: 18]

Type lines are not always two continuous ridges. In fact, they are more often found to be broken. When there is a definite break in a type line, the ridge immediately *outside* of it is considered as its continuation, as shown by the emphasized ridges in figure 13.

Sometimes type lines may be very short. Care must be exercised in their location. Notice the right type line in figure 14.

When locating type lines it is necessary to keep in mind the distinction between a divergence and a bifurcation (fig. 15).

A bifurcation is the forking or dividing of one line into two or more branches.

A divergence is the spreading apart of two lines which have been running parallel or nearly parallel.

According to the narrow meaning of the words in fingerprint parlance, a single ridge may bifurcate, but it may not be said to diverge. Therefore, with one exception, the two forks of a bifurcation may never constitute type lines. The exception is when the forks run parallel after bifurcating and then diverge. In such a case the two forks become the two innermost ridges required by the definition. In illustration 16, the ridges marked "A--A" are type lines even though they proceed from a bifurcation. In figure 17, however, the ridges A--A are not the type lines because the forks of the bifurcation do not run parallel with each other. Instead, the ridges marked "T" are the type lines.

Angles are never formed by a single ridge but by the abutting of one ridge against another. Therefore, an angular formation cannot be used as a type line. In figure 18, ridges A and B join at an angle. Ridge B does not run parallel with ridge D; ridge A does not diverge. Ridges C and D, therefore, are the type lines.

Focal points--Within the pattern areas of loops and whorls are enclosed the focal points which are used to classify them. These points are called delta and core.

The delta is that point on a ridge at or in front of and nearest the center of the divergence of the type lines.

It may be:

- A bifurcation

- An abrupt ending ridge

- A dot

- A short ridge

- A meeting of two ridges

- A point on the first recurving ridge located nearest to the center and in front of the divergence of the type lines.

The concept of the delta may perhaps be clarified by further exposition. Webster furnishes the following definition: "(1) Delta is the name of the fourth letter of the Greek alphabet (equivalent to the English D) from the Phoenician name for the corresponding letter. The Greeks called the alluvial deposit at the mouth of the Nile, from its shape, the Delta of the Nile. (2) A tract of land shaped like the letter "delta," especially when the land is alluvial, and enclosed within two or more mouths of a river, as the Delta of the Ganges, of the Nile, of the

Mississippi" (fig. 19).

When the use of the word "delta" in physical geography is fully grasped, its fitness as applied in fingerprint work will become evident. Rivers wear away their banks and carry them along in their waters in the form of a fine sediment. As the rivers unite with seas or lakes, the onward sweep of the water is lessened, and the sediment, becoming comparatively still, sinks to the bottom where there is formed a shoal which gradually grows, as more and more is precipitated, until at length a portion of the shoal becomes higher than the ordinary level of the stream. There is a similarity between the use of the word "delta" in physical geography and in fingerprints. The island formed in front of the diverging sides of the banks where the stream empties at its mouth corresponds to the delta in fingerprints, which is the first obstruction of any nature at the point of divergence of the type lines in front of or nearest the center of the divergence.

[Illustration: 19]

[Illustration: 20]

In figure 20, the dot marked "delta" is considered as the delta because it is the first ridge or part of a ridge nearest the point of divergence of the two type lines. If the dot were not present, point B on ridge C, as shown in the figure, would be considered as the delta. This would be equally true whether the ridges were connected with one of the type lines, both type lines, or disconnected altogether. In figure 20, with the dot as the delta, the first ridge count is ridge C. If the dot were not present, point B on ridge C would be considered as the delta and the first count would be ridge D. The lines X--X and Y--Y are the type lines, not X--A and Y--Z.

In figures 21 to 24, the heavy lines A--A and B--B are type lines with the delta at point D.

[Illustration: 21]

[Illustration: 22]

CHAPTER II

[Illustration: 23]

[Illustration: 24]

[Illustration: 25]

[Illustration: 26]

Figure 25 shows ridge A bifurcating from the lower type line inside the pattern area. Bifurcations are also present within this pattern at points B and C. The bifurcation at the point marked "delta" is the only one which fulfills all conditions necessary for its location. It should be understood that the diverging type lines must be present in all delta formations and that wherever one of the formations mentioned in the definition of a delta may be, it must be located midway between two diverging type lines at or just in front of where they diverge in order to satisfy the definition and qualify as a delta.

When there is a choice between two or more possible deltas, the following rules govern:

- *The delta may not be located at a bifurcation which does not open toward the core.*

In figure 26, the bifurcation at E is closer to the core than the bifurcation at D. However, E is not immediately in front of the divergence of the type lines and it *does not* open toward the core. A--A and B--B are the only possible type lines in this sketch and it follows, therefore, that the bifurcation at D must be called the delta. The first ridge count would be ridge C.

- *When there is a choice between a bifurcation and another type of delta, the bifurcation is selected.*

A problem of this type is shown in figure 27. The dot, A, and the bifurcation are equally close to the divergence of the type lines, but the bifurcation is selected as the delta. The ridges marked "T" are the type lines.

[Illustration: 27]

[Illustration: 28]

- When there are two or more possible deltas which conform to the definition, the one nearest the core is chosen.

Prints are sometimes found wherein a single ridge enters the pattern area with two or more bifurcations opening toward the core. Figure 28 is an example of this. Ridge A enters the pattern area and bifurcates at points X and D. The bifurcation at D, which is the closer to the core, is the delta and conforms to the rule for deltas. A--A and B--B are the type lines. A bifurcation which does not conform to the definition should not be considered as a delta irrespective of its distance from the core.

- The delta may not be located in the middle of a ridge running between the type lines toward the core, but at the nearer end only.

The location of the delta in this case depends entirely upon the point of origin of the ridge running between the type lines toward the core. If the ridge is entirely within the pattern area, the delta is located at the end nearer the point of divergence of the type lines. Figure 29 is an example of this kind.

[Illustration: 29]

[Illustration: 30]

If the ridge enters the pattern area from a point below the divergence of the type lines, however, the delta must be located at the end nearer the core. Ridge A in figure 30 is of this type.

In figure 31, A--A and B--B are the type lines, with the dot as the delta. The bifurcations cannot be considered as they do not open toward the core.

[Illustration: 31]

CHAPTER II
18

[Illustration: 32]

In figure 32, the dot cannot be the delta because line D cannot be considered as a type line. It does not run parallel to type line A--A at any point. The same reason prevents line E from being a type line. The end of ridge E is the only possible delta as it is a point on the ridge nearest to the center of divergence of the type lines. The other type line is, of course, B--B.

The delta is the point from which to start in ridge counting. In the loop type pattern the ridges intervening between the delta and the core are counted. The core is the second of the two focal points.

The core, as the name implies, is the approximate center of the finger impression. It will be necessary to concern ourselves with the core of the loop type only. The following rules govern the selection of the core of a loop:

- *The core is placed upon or within the innermost sufficient recurve.*

- *When the innermost sufficient recurve contains no ending ridge or rod rising as high as the shoulders of the loop, the core is placed on the shoulder of the loop farther from the delta.*

- *When the innermost sufficient recurve contains an uneven number of rods rising as high as the shoulders, the core is placed upon the end of the center rod whether it touches the looping ridge or not.*

- *When the innermost sufficient recurve contains an even number of rods rising as high as the shoulders, the core is placed upon the end of the farther one of the two center rods, the two center rods being treated as though they were connected by a recurving ridge.*

The shoulders of a loop are the points at which the recurving ridge definitely turns inward or curves.

Figures 33 to 38 reflect the focal points of a series of loops. In figure 39, there are two rods, but the rod marked "A" does not rise as high as the shoulder line X, so the core is at B.

[Illustration: 33]

[Illustration: 34]

Figures 40 to 45 illustrate the rule that a recurve must have no appendage abutting upon it at a right angle between the shoulders and on the outside. If such an appendage is present between the shoulders of a loop, that loop is considered spoiled and the next loop outside will be considered to locate the core. In each of the figures, the point C indicates the core. Appendages will be further explained in the section concerning loops.

[Illustration: 35]

[Illustration: 36]

[Illustration: 37]

[Illustration: 38]

[Illustration: 39]

[Illustration: 40]

Figures 46 to 48 reflect interlocking loops at the center, while figure 49 has two loops side by side at the center. In all these cases the two loops are considered as one. In figure 46, when the shoulder line X--X is drawn it is found to cross exactly at the point of intersection of the two loops. The two loops are considered one, with one rod, the core being placed at C. In figure 47, the shoulder line X--X is above the point of intersection of the two loops. The two are considered as one, with two rods, the core being at C. In figure 48, the shoulder line X--X is below the point of intersection of the loops. Again the two are treated as one, with two rods, the core being placed at C. In figure 49, the two are treated as one, with two rods, the core being placed at C.

[Illustration: 41]

[Illustration: 42]

CHAPTER II

[Illustration: 43]

[Illustration: 44]

[Illustration: 45]

[Illustration: 46]

[Illustration: 47]

[Illustration: 48]

[Illustration: 49]

[Illustration: 50]

[Illustration: 51]

[Illustration: 52]

In figure 50, the delta is formed by a bifurcation which is not connected with either of the type lines. The first ridge count in this instance is ridge C. If the bifurcation were not present, the delta would be a point on ridge C and the first ridge count would be ridge D. In figure 51, the ridge which bifurcates is connected with the lower type line. The delta in this would be located on the bifurcation as designated and the first ridge count would be ridge C. Figure 52 reflects the same type of delta shown in the previous figure in that the ridge is bifurcating from a type line and then bifurcates again to form the delta.

A white space must intervene between the delta and the first ridge count. If no such interval exists, the first ridge must be disregarded. In figures 53 and 54, the first ridge beyond the delta is counted. In figure 55, it is not counted because there is no interval between it and the delta. Notice that the ridge running from the delta toward the core is in a straight line between them. If it were not, of course, an interval would intervene as in figures 53 and 54.

[Illustration: 53]

[Illustration: 54]

[Illustration: 55]

The loop

In fingerprints, as well as in the usual application of the word "loop," there cannot be a loop unless there is a recurve or turning back on itself of one or more of the ridges. Other conditions have to be considered, however. A pattern must possess several requisites before it may be properly classified as a loop. This type of pattern is the most numerous of all and constitutes about 65 percent of all prints.

A loop is that type of fingerprint pattern in which one or more of the ridges enter on either side of the impression, recurve, touch or pass an imaginary line drawn from the delta to the core, and terminate or tend to terminate on or toward the same side of the impression from whence such ridge or ridges entered.

Essentials of a loop

- A sufficient recurve.

- A delta.

- A ridge count across a looping ridge.

A sufficient recurve may be defined as that part of a recurving ridge between the shoulders of a loop. It must be free of any appendages abutting upon the outside of the recurve at a right angle.

Appendages--Some explanation is necessary of the importance attached to appendages. Much care must be exercised in interpreting appendages because they sometimes change the shape of the recurving ridge to which they are connected. For example, a loop with an appendage abutting upon its recurve between the shoulders and at right angles, as in illustration 56, will appear sometimes as in illustration 57 with the recurve totally destroyed. For further examples see figures 161 to 184.

CHAPTER II

[Illustration: 56]

[Illustration: 57]

The same is true of a whorl recurve, as in figures 58 and 59.

It is necessary, therefore, to consider and classify figures 56 and 58 as if they actually appeared as in figures 57 and 59.

In figure 60, there is a ridge marked "A" which enters on one side of the impression and, after recurving, passes an imaginary line drawn from the core C to delta D, and terminates on the same side of the impression from which it entered, marked "B", thus fulfilling all the conditions required in the definition of a loop. X and Y are the type lines. It will be noted in figure 61 that there is a ridge which enters on one side of the impression, recurves, and passes an imaginary line drawn from the delta to the core. It does not terminate on the side from which it entered but has a tendency to do so. In this case, all the requirements of the loop have been met, and consequently it is classified as such.

[Illustration: 58]

[Illustration: 59]

[Illustration: 60]

Figure 62 shows a ridge entering on one side of the impression, recurving, and passing beyond an imaginary line drawn from the delta to the core, although opposite from the pattern shown in figure 61. After passing the imaginary line, the recurving ridge does not terminate on the side of the impression from which it entered, but it has a tendency to do so, and the pattern is, therefore, a loop.

In figure 63, a ridge enters on one side of the impression and then recurves, containing two rods within it, each of which rises as high as the shoulder of the loop. From our study of cores, we know that the top of the rod more distant from the delta is the core, but the recurving ridge does not pass the imaginary line. For that reason the pattern is

CHAPTER II

not classified as a loop, but is given the preferential classification of a tented arch due to the lack of one of the loop requisites. The proper location of the core and delta is of extreme importance, for an error in the location of either might cause this pattern to be classified as a loop.

Figure 64 reflects a similar condition.

[Illustration: 61]

[Illustration: 62]

[Illustration: 63]

[Illustration: 64]

[Illustration: 65]

[Illustration: 66]

In figure 65, there is a looping ridge A which enters on one side of the impression. The ridges B and C are the type lines. As determined by rules already stated, the location of the core and the location of the delta are shown, and if an imaginary line were placed on the core and delta, the recurving ridge A would cross it. This is another figure showing a ridge which does not terminate on the side of the impression from which it entered but tends to do so, and, therefore, is considered as a loop.

In figure 66, we have a print which is similar in many respects to the one described in the preceding paragraph, but here the recurving ridge A continues and tends to terminate on the *opposite* side of the impression from which it entered. For this reason the pattern is not a loop, but a tented arch. The recurving ridge must touch or pass the imaginary line between delta and core and at least tend to pass out toward the side from which it entered, so that a ridge count of at least one can be obtained.

[Illustration: 67]

Figure 67 shows a ridge which enters on one side of the impression and, after flowing toward the center, turns or loops on itself and terminates on the same side from whence it entered. This pattern would be classified as a loop. This pattern should be distinguished from the pattern appearing in figure 139. Careful study of the pattern in figure 67 reveals that the core is located at C and the delta D. The imaginary line between these points will be crossed by the ridge forming a loop. In figure 139, the core is located on the recurve and an imaginary line between the delta and the core does not cross a looping ridge. Figure 139 is thus classified as a tented arch, as will be seen later.

Figure 68 shows at the center of the print a ridge which forms a pocket. It will be noticed that ridge A does not begin on the edge of the print, but this is of no significance. The ridge A within the pattern area recurves or loops, passing the imaginary line between the delta and the core, and tends to terminate toward the same side of the impression from whence it entered. This is a loop pattern possessing all of the requirements.

In figures 69 and 70, it will be observed that there is a ridge entering on one side of the pattern which recurves and then turns back on itself. These patterns are different from any others which have been shown in this respect but are classified as loops. In each of the patterns the core and delta are marked "C" and "D". The reader should trace the type lines in order to ascertain why the delta is located at point D, and then apply the delta rule.

[Illustration: 68]

[Illustration: 69]

Figure 71 is an example of loops as they appear on the rolled impression portion of a fingerprint card.

[Illustration: 70]

[Illustration: 71]

CHAPTER II

Right Hand ------- 1. Thumb | 2. Index | 3. Middle | 4. Ring | 5. Little | finger | finger | finger | finger [Illustration]|[Illustration]|[Illustration]|[Illustration]|[Illustration]
-------- Left Hand
-------- 6. Thumb | 7. Index | 8. Middle | 9. Ring | 10. Little | finger | finger | finger | finger [Illustration]|[Illustration]|[Illustration]|[Illustration]|[Illustration]

Ridge counting

The number of ridges intervening between the delta and the core is known as the ridge count. The technical employees of the Federal Bureau of Investigation count each ridge which *crosses or touches* an imaginary line drawn from the delta to the core. Neither delta nor core is counted. A red line upon the reticule of the fingerprint glass is used to insure absolute accuracy. In the event there is a bifurcation of a ridge exactly at the point where the imaginary line would be drawn, two ridges are counted. Where the line crosses an island, both sides are counted. Fragments and dots are counted as ridges only if they appear to be as thick and heavy as the other ridges in the immediate pattern. Variations in inking and pressure must, of course, be considered.

Figures 72 to 97 and figures 98 to 101 show various loop patterns. The reader should examine each one carefully in order to study the cores and deltas and to verify the count which has been placed below each pattern.

[Illustration: 72. 12 counts.]

[Illustration: 73. 2 counts.]

[Illustration: 74. 16 counts.]

[Illustration: 75. 7 counts.]

[Illustration: 76. 4 counts.]

[Illustration: 77. 7 counts.]

CHAPTER II

[Illustration: 78. 15 counts.]

[Illustration: 79. 16 counts.]

[Illustration: 80. 9 counts.]

[Illustration: 81. 3 counts.]

[Illustration: 82. 9 counts.]

[Illustration: 83. 20 counts.]

[Illustration: 84. 6 counts.]

[Illustration: 85. 2 counts.]

[Illustration: 86. 8 counts.]

[Illustration: 87. 14 counts.]

[Illustration: 88. 5 counts.]

[Illustration: 89. 12 counts.]

[Illustration: 90. 12 counts.]

[Illustration: 91. 3 counts.]

[Illustration: 92. 16 counts.]

[Illustration: 93. 14 counts.]

[Illustration: 94. 16 counts.]

[Illustration: 95. 18 counts.]

[Illustration: 96. 2 counts.]

[Illustration: 97. 1 count.]

[Illustration: 98. 1 count.]

[Illustration: 99. 2 counts.]

[Illustration: 100. 8 counts.]

[Illustration: 101. 13 counts.]

Figure 102 is a sketch reflecting the various types of ridges which the classifier will encounter when engaging in counting loop patterns.

In figure 103, the lighter lines are caused by the splitting or fraying of the ridges. Sometimes ingrained dirt will cause a similar condition between the ridges. These lines are not considered ridges and should not be counted.

In figure 104, the dot is not the delta because it is not as thick and heavy as the other ridges and might not be present if the finger were not perfectly inked and printed.

When the core is located on a spike which touches the inside of the innermost recurving ridge, the recurve is included in the ridge count only when the delta is located below a line drawn at right angles to the spike.

Figures 105 and 106 are examples of this rule.

If the delta is located in areas A, the recurving ridge is counted.

If the delta is located in areas B, the recurving ridge is not counted.

[Illustration: 102]

LOOP 25 RIDGE COUNTS

1. SHORT RIDGE 2. } 3. } BIFURCATION 4. } 5. } BIFURCATION 6. RIDGE 7. ENDING RIDGE 8. } 9. } BIFURCATION 10. RIDGE 11. ENDING RIDGE 12. RIDGE 13. SHORT RIDGE 14. } 15. } BIFURCATION 16. } 17. } ISLAND 18. } 19. } BIFURCATION 20.

CHAPTER II

ENDING RIDGE 21. DOT 22. RIDGE 23. } 24. } ISLAND 25. ENDING RIDGE

[Illustration: 103]

[Illustration: 104]

[Illustration: 105]

[Illustration: 106]

Radial and ulnar loops

The terms "radial" and "ulnar" are derived from the radius and ulna bones of the forearm. Loops which flow in the direction of the ulna bone (toward the little finger) are called ulnar loops and those which flow in the direction of the radius bone are called radial loops.

For test purposes, fingers of the right hand may be placed on the corresponding print of the right hand appearing in figure 71, and it will be noticed that the side of each finger which is nearer to the thumb on the hand is also nearer to the thumb on the fingerprint card. Place the fingers of the *left* hand on the corresponding prints of the *left* hand shown in figure 71. It will be noticed that the arrangement of the prints on the card is the *reverse* of the arrangement of the fingers on the hand. *The classification of loops is based on the way the loops flow on the hand (not the card), so that on the fingerprint card for the left hand, loops flowing toward the thumb impression are ulnar, and loops flowing toward the little finger impression are radial.*

The plain arch

In plain arches the ridges enter on one side of the impression and flow or tend to flow out the other with a rise or wave in the center. The plain arch is the most simple of all fingerprint patterns, and it is easily distinguished. Figures 107 to 118 are examples of the plain arch. It will be noted that there may be various ridge formations such as ending ridges, bifurcations, dots and islands involved in this type of pattern, but they all tend to follow the general ridge contour; i.e., they enter on one side, make a rise or wave in the center, and flow or tend to flow out the other side.

[Illustration: 107]

[Illustration: 108]

[Illustration: 109]

[Illustration: 110]

Figures 119 and 120 are examples of plain arches which approximate tented arches. Also, figure 121 is a plain arch approximating a tented arch as the rising ridge cannot be considered an upthrust because it is a continuous, and not an ending, ridge. (See following explanation of the tented arch.)

[Illustration: 111]

[Illustration: 112]

[Illustration: 113]

[Illustration: 114]

[Illustration: 115]

[Illustration: 116]

[Illustration: 117]

[Illustration: 118]

[Illustration: 119]

[Illustration: 120]

[Illustration: 121]

CHAPTER II

The tented arch

In the tented arch, most of the ridges enter upon one side of the impression and flow or tend to flow out upon the other side, as in the plain arch type; however, the ridge or ridges at the center do not. There are three types of tented arches:

- The type in which ridges at the center form a definite angle; i.e., 90° or less.

- The type in which one or more ridges at the center form an upthrust. An upthrust is an ending ridge of any length rising at a sufficient degree from the horizontal plane; i.e., 45° or more.

- The type approaching the loop type, possessing two of the basic or essential characteristics of the loop, but lacking the third.

Figures 122 to 133 are examples of the tented arch.

[Illustration: 122]

[Illustration: 123]

[Illustration: 124]

[Illustration: 125]

[Illustration: 126]

[Illustration: 127]

[Illustration: 128]

[Illustration: 129]

[Illustration: 130]

[Illustration: 131]

[Illustration: 132]

[Illustration: 133]

Figures 122 to 124 are of the type possessing an angle.

Figures 125 to 129 reflect the type possessing an upthrust.

Figures 130 to 133 show the type approaching the loop but lacking one characteristic.

Tented arches and some forms of the loop are often confused. It should be remembered by the reader that the *mere converging of two ridges does not form a recurve, without which there can be no loop*. On the other hand, there are many patterns which at first sight resemble tented arches but which on close inspection are found to be loops, as where one looping ridge will be found in an almost vertical position within the pattern area, entirely free from and passing in front of the delta.

Figure 134 is a tented arch. The ridge marked "A--A" in the sketch enters on one side of the impression and flows to the other with an acute rise in the center. Ridge C strikes into A at point B and should not be considered as a bifurcating ridge. The ridges marked "D--D" would form a tented arch if the rest of the pattern were absent.

[Illustration: 134]

[Illustration: 135]

Figure 135 is a sketch of a pattern reflecting a ridge, A--B, entering on one side of the impression, recurving, and making its exit on the other side of the impression. The reader should study this sketch carefully. It should be borne in mind that there must be a ridge entering on one side of the impression and recurving in order to make its exit on the same side from which it entered, or having a tendency to make its exit on that side, before a pattern can be considered for possible classification as a loop. This pattern is a tented arch of the upthrust type. The upthrust is C. There is also an angle at E. D cannot be termed as a delta, as the ridge to the left of D cannot be considered a type line because it does not diverge from the ridge to the right of D but turns and goes in the same direction.

In connection with the types of tented arches, the reader is referred to the third type. This form of tented arch, the one which approaches the loop, may have *any combination of two of the three basic loop characteristics, lacking the third*. These three loop characteristics are, to repeat:

- *A sufficient recurve.*

- *A delta.*

- *A ridge count across a looping ridge.*

It must be remembered that a recurve must be free of any appendage abutting upon it at a right angle between the shoulders, and a true ridge count is obtained only by crossing a looping ridge freely, with a white space intervening between the delta and the ridge to be counted.

[Illustration: 136]

[Illustration: 137]

Figures 136 and 137 are tented arches having loop formations within the pattern area but with deltas upon the loops, by reason of which it is impossible to secure a ridge count. The type lines run parallel from the left in figures 136 and 137. These tented arches have two of the loop characteristics, recurve and delta, but lack the third, the ridge count.

In figure 138, the reader will note the similarity to the figures 136 and 137. The only difference is that in this figure the type lines are running parallel from the right. It will be noted from these three patterns that the spaces between the type lines at their divergence show nothing which could be considered as delta formations except the looping ridges. Such patterns are classified as tented arches because the ridge count necessary for a loop is lacking.

[Illustration: 138]

[Illustration: 139]

[Illustration: 140]

[Illustration: 141]

Figure 139 is an example of a tented arch. In this pattern, if the looping ridge approached the vertical it could possibly be a one-count loop. Once studied, however, the pattern presents no real difficulty. There are no ridges intervening between the delta, which is formed by a bifurcation, and the core. It will be noted that the core, in this case, is at the center of the recurve, unlike those loops which are broadside to the delta and in which the core is placed upon the shoulder. This pattern has a recurve and a separate delta, but it still lacks the ridge count necessary to make it a loop.

Figures 140 and 141 are examples of tented arches. These two figures are similar in many ways. Each of these prints has three abrupt ending ridges but lacks a recurve; however, in figure 141 a delta is present in addition to the three abrupt ending ridges. This condition does not exist in figure 140, where the lower ending ridge is the delta.

When interpreting a pattern consisting of two ending ridges and a delta but lacking a recurve, do not confuse the ridge count of the tented arch with that of the ridge count for the loop. The ridge count of the tented arch is merely a convention of fingerprinting, a fiction designed to facilitate a scientific classification of tented arches, and has no connection with a loop. To obtain a true ridge count there must be a looping ridge which is crossed freely by an imaginary line drawn between the delta and the core. The ridge count referred to as such in connection with the tented arches possessing ending ridges and no recurve is obtained by imagining that the ending ridges are joined by a recurve only for the purpose of locating the core and obtaining a ridge count. If this point is secure in the mind of the classifier, little difficulty will be encountered.

CHAPTER II   34

Figures 140 and 141, then, are tented arches because they have two of the characteristics of a loop, delta and ridge count, but lack the third, the recurve.

[Illustration: 142]

[Illustration: 143]

Figure 142 is a loop formation connected with the delta but having no ridge count across a looping ridge. By drawing an imaginary line from the core, which is at the top of the rod in the center of the pattern, to the delta, it will be noted that there is no recurving ridge passing between this rod and the delta; and, therefore, no ridge count can result. This pattern is classified as a tented arch. There must be a white space between the delta and the first ridge counted, or it may not be counted. Figure 143 is also a tented arch because no ridge count across a looping ridge can be obtained, the bifurcations being connected to each other and to the loop in a straight line between delta and core. The looping ridge is not crossed freely. No white space intervenes between the delta and the loop. These patterns are tented arches because they possess two of the characteristics of a loop, a delta and a recurve, but lack the third, a ridge count across a looping ridge.

Figure 144 is a tented arch combining two of the types. There is an angle formed by ridge *a* abutting upon ridge *b*. There are also the elements of the type approaching a loop, as it has a delta and ridge count but lacks a recurve.

[Illustration: 144]

[Illustration: 145]

[Illustration: 146]

Figures 145 to 148 are tented arches because of the angles formed by the abutting ridges at the center of the patterns.

CHAPTER II

Figure 149 is a tented arch because of the upthrust present at the center of the pattern. The presence of the slightest upthrust at the center of the impression is enough to make a pattern a tented arch.

[Illustration: 147]

[Illustration: 148]

[Illustration: 149]

[Illustration: 150]

An upthrust must be an ending ridge. If continuous as in figure 150, no angle being present, the pattern is classified as a plain arch.

Figures 151 to 153 are plain arches. Figure 154 is a tented arch.

Figure 155 is a plain arch because it is readily seen that the apparent upthrust A is a continuation of the curving ridge B. Figure 156 is a tented arch because ridge A is an independent upthrust, and not a continuation of ridge B.

[Illustration: 151]

[Illustration: 152]

[Illustration: 153]

[Illustration: 154]

[Illustration: 155]

[Illustration: 156]

Figures 157 and 158 are plain arches. Figure 158 cannot be said to be a looping ridge, because by definition a loop must pass out or tend to pass out upon the side from which it entered. This apparent loop passes out upon the opposite side and cannot be said to tend to flow out upon the same side.

[Illustration: 157]

[Illustration: 158]

In figures 159 and 160, there are ending ridges rising at about the same degree from the horizontal plane.

Figure 159, however, is a plain arch, while 160 is a tented arch. This differentiation is necessary because, if the first pattern were printed crookedly upon the fingerprint card so that the ending ridge was nearer the horizontal plane, there would be no way to ascertain the true horizontal plane of the pattern (if the fissure of the finger did not appear). In other words, there would be no means of knowing that there was sufficient rise to be called an upthrust, so that it is safe to classify the print as a plain arch only. In figure 160, however, no matter how it is printed, the presence of a sufficient rise could always be ascertained because of the space intervening between the ending ridge and the ridge immediately beneath it, so that it is safe to classify such a pattern as a tented arch. The test is, *if the ridges on both sides of the ending ridge follow its direction or flow trend, the print may be classified as a plain arch. If, however, the ridges on only one side follow its direction, the print is a tented arch.*

[Illustration: 159]

[Illustration: 160]

An upthrust, then, must not only be an ending ridge rising at a sufficient degree from the horizontal plane, but there must also be a space between the ending ridge and the ridge immediately beneath it. *This, however, is not necessary for a short upthrust or spike, or any upthrust which rises perpendicularly.*

In connection with the proper classification to be assigned to those borderline loop-tented arch cases where an appendage or spike is thrusting out from the recurve, it is necessary to remember that *an appendage or a spike abutting upon a recurve at right angles in the space between the shoulders of a loop on the outside is considered to spoil the recurve.*

CHAPTER II

If the appending ridge flows off the looping ridge smoothly in such a way that it forms a bifurcation and not an abutment of two ridges at a right angle, the recurve is considered as remaining intact. The test is to trace the looping ridge toward the appendage, and if, when it is reached, the tracing may be continued as readily upon the appendage as upon the looping ridge, with no sudden, sharp change of direction, the recurve is sufficient. Figures 161 to 184 should be studied with this in mind.

[Illustration: 161. Tented arch.]

[Illustration: 162. Tented arch.]

[Illustration: 163. Tented arch.]

[Illustration: 164. Tented arch.]

[Illustration: 165. Tented arch.]

[Illustration: 166. Tented arch.]

[Illustration: 167. Tented arch.]

[Illustration: 168. Tented arch.]

[Illustration: 169. Loop.]

[Illustration: 170. Loop.]

[Illustration: 171. Loop.]

[Illustration: 172. Loop.]

[Illustration: 173. Loop.]

[Illustration: 174. Loop.]

[Illustration: 175. Loop.]

CHAPTER II   38

[Illustration: 176. Tented arch.]

[Illustration: 177. Tented arch.]

[Illustration: 178. Tented arch.]

[Illustration: 179. Loop.]

[Illustration: 180. Loop.]

[Illustration: 181. Loop.]

[Illustration: 182. Loop.]

[Illustration: 183. Loop.]

[Illustration: 184. Loop.]

Figures 185 to 190 show additional examples of tented arches.

[Illustration: 185]

[Illustration: 186]

[Illustration: 187]

[Illustration: 188]

[Illustration: 189]

[Illustration: 190]

The reason that figure 185 is given the classification of a tented arch is because of the presence of all the loop requirements with the exception of one, which is the recurve. In this pattern appear three ending ridges. The lowest ending ridge provides the delta, and the other two by the convention explained previously, provide the ridge count. It is a tented arch, then, of the type approaching the loop, with two of the characteristics, but lacking the third, a recurve. Figures 186 and 187 are tented arches of the same type. A close examination of these prints will reveal that when the imaginary line is drawn between delta and core no ridge count across a looping ridge can be obtained. It must be remembered that the core of a loop may not be placed below the shoulder line. Lacking one of the three characteristics of a loop, these patterns must be classified as tented arches. When figure 188 is examined, it will be noticed that the recurve is spoiled by the appendage abutting upon it between the shoulders at a right angle, so it must also be classified with the tented arches. In figure 189, the only possible delta must be placed upon the looping ridge, thus preventing a ridge count although delta and recurve are present. Figure 190 is assigned the classification of a tented arch. One of the requirements of a loop type is that the ridge enters on one side, recurves, and makes its exit on the side from which it entered. This, of course, makes it necessary that the ridge pass between the delta and the core. It will be noted from this figure that although this ridge passes between the delta and the core, it does not show any tendency to make its exit on the side from which it entered, and therefore the loop classification is precluded, and it is a tented arch.

The whorl

The patterns to which numerical values are assigned in deriving the "primary" in the extension of the Henry System of fingerprint classification used by the Federal Bureau of Investigation are the whorl-type patterns, which occur in about 30 percent of all fingerprints.

The whorl is that type of pattern in which at least two deltas are present with a recurve in front in each. Figures 191 to 193 reflect the minimum requirements for the whorl.

[Illustration: 191]

[Illustration: 192]

[Illustration: 193]

It is important to note that the above definition is very general; however, this pattern may be subdivided for extension purposes in

CHAPTER II

large groups where whorls are predominant. Even though this extension may be used, all types of whorls are grouped together under the general classification of "Whorl" and are designated by the letter "W".

The aforementioned subdivisions are as follows: The Plain Whorl, The Central Pocket Loop, The Double Loop, and The Accidental.

The plain whorl

The "plain whorl" consists of the simplest form of whorl construction and is the most common of the whorl subdivisions. It is designated by the symbol "W" for both general classification and extension purposes.

The plain whorl has two deltas and at least one ridge making a complete circuit, which may be spiral, oval, circular, or any variant of a circle. An imaginary line drawn between the two deltas must touch or cross at least one of the recurving ridges within the inner pattern area. A recurving ridge, however, which has an appendage connected with it in the line of flow cannot be construed as a circuit. An appendage connected at that point is considered to spoil the recurve on that side.

Figures 194 to 211 are typical examples of the plain whorl type. Figure 212 is, however, a loop, as the circuit is spoiled on one side by an appendage.

[Illustration: 194]

[Illustration: 195]

[Illustration: 196]

[Illustration: 197]

[Illustration: 198]

[Illustration: 199]

[Illustration: 200]

CHAPTER II

[Illustration: 201]

[Illustration: 202]

[Illustration: 203]

[Illustration: 204]

[Illustration: 205]

[Illustration: 206]

[Illustration: 207]

[Illustration: 208]

[Illustration: 209]

[Illustration: 210]

[Illustration: 211]

[Illustration: 212]

Central pocket loop

Within the whorl group, the subclassification type "central pocket loop" is used for extension purposes only. In general classification it is designated by the letter "W". Figures 213 to 236 are central pocket loops.

The central pocket loop type of whorl has two deltas and at least one ridge making a complete circuit, which may be spiral, oval, circular, or any variant of a circle. An imaginary line drawn between the two deltas must not touch or cross any of the recurving ridges within the inner pattern area. A recurving ridge, however, which has an appendage connected with it in the line of flow and on the delta side cannot be construed as a circuit. An appendage connected at that point is considered to spoil the recurve on that side.

CHAPTER II

In lieu of a recurve in front of the delta in the inner pattern area, an obstruction at right angles to the line of flow will suffice.

It is necessary that the inner line of flow be fixed artificially. *The inner line of flow is determined by drawing an imaginary line between the inner delta and the center of the innermost recurve or looping ridge.*

In the central pocket loop, one or more of the simple recurves of the plain loop type usually recurve a second time to form a pocket within the loop. The second recurve, however, need not be a continuation of--or even connected with--the first. It may be an independent ridge.

If no second recurve is present, an obstruction at right angles to the inner line of flow is acceptable in lieu of it. An obstruction may be either curved or straight. A dot, of course, may not be considered an obstruction.

The definition does not require a recurve to cross the line of flow at right angles. The angle test needs to be applied to obstructions only.

The recurve or obstruction of the central pocket loop, as that of the plain whorl, must be free of any appendage connected to it at the point crossed by the line of flow and on the delta side. An appendage at that point is considered to spoil the recurve or obstruction.

[Illustration: 213]

[Illustration: 214]

[Illustration: 215]

[Illustration: 216]

[Illustration: 217]

[Illustration: 218]

[Illustration: 219]

CHAPTER II

[Illustration: 220]

[Illustration: 221]

[Illustration: 222]

[Illustration: 223]

[Illustration: 224]

[Illustration: 225]

[Illustration: 226]

[Illustration: 227]

[Illustration: 228]

[Illustration: 229]

[Illustration: 230]

[Illustration: 231]

[Illustration: 232]

[Illustration: 233]

[Illustration: 234]

[Illustration: 235]

[Illustration: 236]

Figures 237 and 238 are also central pocket loops despite the appendages connected to the recurves, because they are not connected at the point crossed by the line of flow.

CHAPTER II

Figure 239, although possessing a recurve, is classified as a loop because the second delta is located on the only recurring ridge.

[Illustration: 237]

[Illustration: 238]

[Illustration: 239]

[Illustration: 240]

[Illustration: 241]

[Illustration: 242]

Figures 240 to 244, although possessing one delta and a delta formation, are classified as loops because the obstructions do not cross the line of flow at right angles.

[Illustration: 243]

[Illustration: 244]

[Illustration: 245]

[Illustration: 246]

[Illustration: 247]

[Illustration: 248]

Figures 245 to 254 have two deltas and one or more recurves, but they are classified as loops because each recurve is spoiled by an appendage connected to it at the point crossed by the line of flow.

[Illustration: 249]

[Illustration: 250]

CHAPTER II   45

[Illustration: 251]

[Illustration: 252]

[Illustration: 253]

[Illustration: 254]

Double loop

Within the whorl group, the subclassification type "double loop" is used for extension purposes only. In general classification it is designated by the letter "W".

The double loop consists of two separate loop formations, with two separate and distinct sets of shoulders, and two deltas.

The word "separate," as used here, does not mean unconnected. The two loops may be connected by an appending ridge provided that it does not abut at right angles between the shoulders of the loop formation. The appendage rule for the loop applies also to the double loop. An appendage abutting upon a loop at right angles between the shoulders is considered to spoil the loop, while an appendage which flows off smoothly is considered to leave the recurve intact.

The fact that there must be two separate loop formations eliminates from consideration as a double loop the "S" type core, the interlocking type core, and the formation with one loop inside another.

The loops of a double loop do not have to conform to the requirements of the loop. In other words, no ridge count is necessary.

It is not essential that both sides of a loop be of equal length, nor that the two loops be of the same size. Neither is it material from which side the loops enter.

The distinction between twinned loops and lateral pocket loops made by Henry and adopted by other authors has been abandoned by the Federal Bureau of Investigation because of the difficulty in locating and tracing the loops. Both types have been consolidated under the classification "double loop."

Figures 255 to 266 are double loops.

[Illustration: 255]

[Illustration: 256]

[Illustration: 257]

[Illustration: 258]

[Illustration: 259]

[Illustration: 260]

[Illustration: 261]

[Illustration: 262]

[Illustration: 263]

[Illustration: 264]

[Illustration: 265]

[Illustration: 266]

Figure 267 is a plain whorl. It is not classified as a double loop as one side of one loop forms the side of the other. Figure 268 is a plain loop. It is not a double loop because all of the recurves of the loop on the right are spoiled by appendages.

[Illustration: 267]

[Illustration: 268]

Accidental

CHAPTER II

Within the whorl group the subdivision type "accidental" is used for extension purposes only. In general classification it is designated by the letter "W" and for extension purposes by the letter "X".

The accidental whorl is a pattern consisting of a combination of two different types of pattern, with the exception of the plain arch, with two or more deltas; or a pattern which possesses some of the requirements for two or more different types; or a pattern which conforms to none of the definitions. It may be a combination of loop and tented arch, loop and whorl, loop and central pocket loop, double loop and central pocket loop, or other such combinations. The plain arch is excluded as it is rather the absence of pattern than a pattern. Underneath every pattern there are ridges running from one side to the other, so that if it were not excluded every pattern but the plain arch would be an accidental whorl.

This subclassification also includes those exceedingly unusual patterns which may not be placed by definition into any other classes.

Figures 269 to 271 are accidentals combining a loop with a tented arch. Figures 272 to 276 combine a loop and a plain whorl or central pocket loop. Figure 277 combines a loop and a double loop. Figure 278 combines a loop and a plain arch, so it is classified as a loop. Figure 279 combines a loop and a tented arch.

[Illustration: 269]

[Illustration: 270]

[Illustration: 271]

[Illustration: 272]

[Illustration: 273]

[Illustration: 274]

[Illustration: 275]

CHAPTER II   48

[Illustration: 276]

[Illustration: 277]

Some whorls may be found which contain ridges conforming to more than one of the whorl subdivisions described. In such cases, the order of preference (if any practical distinction need be made) should be: (1) accidental, (2) double loop, (3) central pocket loop, (4) plain whorl.

[Illustration: 278]

[Illustration: 279]

[Illustration: 280]

[Illustration: 281]

Whorl tracing

The technique of whorl tracing depends upon the establishment of the focal points--the deltas. Every whorl has two or more. When the deltas have been located, the ridge emanating from the lower side or point of the extreme left delta is traced until the point nearest or opposite the extreme right delta is reached. The number of ridges intervening between the tracing ridge and the right delta are then counted. If the ridge traced passes inside of (above) the right delta, and three or more ridges intervene between the tracing ridge and the delta, the tracing is designated as an "inner"--I (fig. 280). If the ridge traced passes outside (below) the right delta, and three or more ridges intervene between the tracing ridge and the right delta, the tracing is designated as an "outer"--O (fig. 281). All other tracings are designated as "meeting"--M (figs. 282 to 287).

[Illustration: 282]

[Illustration: 283]

[Illustration: 284]

CHAPTER II 49

[Illustration: 285]

[Illustration: 286]

[Illustration: 287]

Tracing begins from the left delta. In no instance is a tracing to begin on a type line. In figure 288, tracing begins at the short ridge which is the left delta. It is true that inasmuch as the short ridge ends immediately the type line is next followed, but this is only because the type line is the next lower ridge. Its status as a type line is independent and has no bearing on the fact that it is being traced. This point is illustrated further in figure 289. This pattern shows an inner tracing. It will be noted that the delta is at the point on the first recurve nearest to the center of the divergence of the type lines. It will be further noted that tracing begins at the point of delta on the left and continues toward the right, passing inside of the right delta, with three ridges intervening between the tracing ridge and the right delta. This shows the tracing to be an inner tracing. If, in this case, the type line were traced (which would be the incorrect procedure), only two ridges would intervene between the tracing ridge and the right delta, resulting in an erroneous meeting tracing. Figure 290 is another example of the application of this rule. This illustration is also an inner whorl.

When the ridge traced ends abruptly, and it is determined that the ridge definitely ends, the tracing drops down to the point on the next lower ridge immediately beneath the point where the ridge above ends, continuing from there. Figure 291, therefore, is an outer whorl.

[Illustration: 288]

[Illustration: 289]

In this connection it should be noted that the rule for dropping to the next lower line applies only when the ridge *definitely* ends. Short breaks in a ridge which may be due to improper inking, the presence of foreign matter on the ridges, enlarged pores, disease, or worn ridges should not be considered as definite ridge endings. The determination of what constitutes a definite ending will depend, of course, upon the good judgment of the classifier. When the question arises as to whether a break encountered in the ridge tracing is a definite ending, or whether there has been interference with a natural impression, the whole pattern should be examined to ascertain whether such breaks are general throughout the pattern. If they are found to be common, consideration should then be given to the possibility that the break is not a definite ridge ending. Appropriate reference tracing should be done in all such cases.

[Illustration: 290]

[Illustration: 291]

Whenever the ridge traced bifurcates, the rule for tracing requires that the lower limb or branch proceeding from the bifurcation be followed. This is illustrated in 292.

[Illustration: 292]

Accidentals often possess three or more deltas. In tracing them only the extreme deltas are considered, the tracing beginning at the extreme left delta and proceeding toward the extreme right delta, as illustrated in figure 293.

In a double loop or accidental the problem of where to stop tracing is sometimes presented. The rule is, *when the tracing passes inside of the right delta, stop at the nearest point to the right delta on the upward trend*, as in figure 294. If no upward trend is present, continue tracing until a point opposite the right delta, or the delta itself, is reached (figs. 295 and 296).

[Illustration: 293]

[Illustration: 294]

[Illustration: 295]

[Illustration: 296]

CHAPTER III

Questionable Patterns

No matter how definite fingerprint rules and pattern definitions are made, there will always be patterns concerning which there is doubt as to the classification they should be given. The primary reason for this is the fact that probably no two fingerprints will ever appear which are exactly alike. Other reasons are differences in the degree of judgment and interpretation of the individual classifying fingerprints, the difference in the amount of pressure used by the person taking the prints, and the amount or kind of ink used. Nothing can be done about faulty inking or pressure once the prints are taken. The patterns which are questionable merely because they seem to have characteristics of two or more types can be classified by strict adherence to the definitions in deducing a preference. The following section is devoted to such patterns with an explanation of each.

[Illustration: 297]

[Illustration: 298]

Figure 297 has two loop formations. The one on the left, however, has an appendage abutting upon the shoulders of its recurve at a right angle. The left portion of the impression, therefore, is of the tented arch type. The combination of two different types of patterns would be classified in the whorl group (accidental), but this impression has only the one delta. The right portion of the pattern detail contains a true loop which fulfills all the loop requirements, i.e., a sufficient recurve, a delta, and a ridge count across a looping ridge. In the choice existing between a tented arch and a loop, preference is given to the loop classification and this impression would be classified as a loop.

[Illustration: 299]

[Illustration: 300]

[Illustration: 301]

CHAPTER III

[Illustration: 302]

Figure 298, at a glance, seems to fulfill the requirements of a whorl (two deltas and a ridge making a complete circuit). The part of the circuit in front of the right delta, however, cannot be construed as a recurving ridge because of the appendage abutting upon it in the line of flow. This pattern, therefore, is a one-count loop.

Figure 299 is a very difficult and unusual pattern. It has characteristics of three types, the whorl, the loop, and the tented arch. It is given the preference of an accidental type of whorl (loop over a tented arch). This pattern should be referenced both as a loop and as a tented arch.

Figure 300 is shown for the purpose of explaining that in the whorl, as this print is, appendages at the top of the recurve will not spoil or affect the recurve. Hence, the impression is a good whorl of the central pocket loop type and needs no reference.

Figure 301 is classified as a whorl of the double loop type. There are present two distinct loops and two deltas (the right delta is not present as the impression was not rolled sufficiently). The pattern is unusual because the loops are side by side and flowing in the same direction. The tracing is an inner tracing.

Figure 302 should present no difficulty. It is classified as a plain arch for its ridge construction follows the rule of a plain arch, i.e., "enter one side and flow or tend to flow to the other."

Figure 303 is a plain arch. The dot at the center is not elongated enough to be considered an upthrust. A dot, even though as thick and heavy as the surrounding ridges, is not considered for any purpose but ridge counting or fixing a delta.

Figure 304 is a pattern somewhat similar to the previous illustration. As indicated before, dots are considered as ridges only in ridge counting and fixing a delta. This pattern, therefore, must be classified as a plain arch, rather than a tented arch with two ending ridges and a delta formation.

Figure 305, although showing an appendage upon each recurve on the left side, is classified as a whorl of the central pocket loop type, with two deltas and a recurve in front of each. To spoil the recurve of a whorl the appendage must be connected to the recurve at the point of contact with the line of flow.

[Illustration: 303]

[Illustration: 304]

[Illustration: 305]

[Illustration: 306]

In figure 306, the impression has two equally good loop formations. As it has but one delta, it cannot be classified as a whorl of the double loop type nor as a loop since it would be difficult to make a preferential choice between the two looping ridges. It is arbitrarily given the classification of a tented arch.

In figure 307, the difficulty lies in locating the delta. The only ridges answering the definition of type lines (ridges running parallel and then diverging to enclose the pattern area) have three ending ridges between them. The type lines, the delta, and the core are located as indicated. The pattern is classified as a six-count loop.

Figure 308 is classified as a tented arch, although it appears at first glance to be a loop. Closer inspection shows that the looping ridge does not tend to go out the side from which it entered but rather seems to proceed downward ending in an abutment forming a definite angle of 90°.

[Illustration: 307]

[Illustration: 308]

[Illustration: 309]

[Illustration: 310]

[Illustration: 311]

In figure 309, an impression is shown which at first appears to be a loop. Closer inspection will show that one of the elements of the loop type is missing, namely, a ridge count across a looping ridge; for it is to be borne in mind that the recurve of the innermost loop should be free of appendages abutting between the shoulders at right angles. The core, in this illustration, therefore, is placed where the appendage of the innermost loop touches the next ridge which is a good recurve. If an imaginary line is placed between delta and core, it will be seen that there are no intervening ridges; hence, there is no ridge count.

Figure 310 is a pattern which contains two elements of a loop but lacks the third. It is classified as a tented arch. Thus an impression having a delta and a recurve, but not having a ridge count across a looping ridge, falls into this classification.

It will be noticed that although this pattern has the resemblance of a plain arch, the center of the impression actually contains a partially formed loop. A recurving ridge enters from the right side and exits in the same direction. A delta also appears just below the recurve. In attempting to obtain a ridge count, it is seen that the imaginary line drawn between the delta and the core runs directly along the ridge emanating from the former and which is joined onto the side of the recurving ridge. For this reason, no ridge count is possible.

[Illustration: 312]

Figure 311 is a tented arch. There are three loop formations, each one of which is spoiled by an appendage abutting upon its recurve between the shoulders at a right angle. It cannot be classified as an accidental as the patterns are all of the same type, i.e., tented arches. An accidental type of whorl is a combination of two or more *different* types of patterns exclusive of the plain arch.

Figure 312 is a loop. It cannot be classified as a whorl of the double loop type because the formation above the lower loop is too pointed and it also has an appendage abutting upon it at a right angle.

Figure 313 at first glance appears to be a whorl of the double loop type. Upon closer inspection, however, it will be noticed that there are no delta formations other than on the recurves. There are, then, two tented arch formations. As two patterns of the same type cannot form an accidental whorl, the impression must be classified as a tented arch.

[Illustration: 313]

[Illustration: 314]

Figure 314 is an accidental whorl, combining a loop and a tented arch. The tented arch is directly beneath the innermost loop, and is of the upthrust type.

Figure 315 consists of a loop over a dot with an apparent second delta. This pattern must be classified as a loop, as a dot may not be considered an upthrust unless elongated vertically.

[Illustration: 315]

[Illustration: 316]

[Illustration: 317]

Even though a dot may be as thick and heavy as the surrounding ridges, it may be considered only in ridge counting or fixing a delta.

Figure 316 at first glance appears to be an accidental whorl, but on closer inspection it proves to be a loop. Although there are three delta formations present, it should be observed that recurving ridges appear in front of only one (D-1).

Figure 317 has the general appearance of a loop. The looping ridge A, at the center, has an appendage B abutting upon its recurve. The abutment is at right angles and therefore spoils the recurve. The pattern is a tented arch.

CHAPTER III 56

Figure 318 is a tented arch which approaches both the loop and the whorl type patterns. It cannot be considered a whorl, however, as the recurve on the left is spoiled by an appendage (figs. 58 and 59). Nor can it be a loop because there is no ridge count across a looping ridge. The pattern, then, is a tented arch of the type possessing two of the basic characteristics of the loop and lacking the third. The delta and the sufficient recurve are present but the ridge count is missing.

Figure 319 seems at first glance to be a double loop. It will be noted, however, that the inner delta formation would be located upon the only looping ridge of the upper loop formation. Since the delta would be located on the only recurve, this recurving ridge is eliminated from consideration. The pattern is classified as a loop.

Figure 320 is a loop of two counts, with the delta at B. There is a ridge making a complete circuit present, but point A cannot be used as a delta because it answers the definition of a type line. It should be considered a delta only if it presented an angular formation. Placing the delta upon the recurve would spoil that recurve.

[Illustration: 318]

[Illustration: 319]

Figure 321 shows two separate looping ridge formations appearing side by side and upon the same side of the delta. The core in such case is placed upon the nearer shoulder of the farther looping ridge from the delta, the two looping ridges being considered as one loop with two rods rising as high as the shoulder. The ridge count would be four (fig. 49).

Figure 322 is an accidental whorl. It is classified thus because it contains elements of three different patterns, the loop, the double loop, and the accidental. In such case the order of preference governs. The delta at the left is point A. The delta at the right is point C. This point becomes the delta since it is the point nearest the center of the divergence of the type lines. Point B is eliminated from consideration as a delta since type lines may not proceed from a bifurcation unless they flow parallel after the bifurcation and before diverging.

CHAPTER III

[Illustration: 320]

[Illustration: 321]

[Illustration: 322]

Figure 323 is a loop. There are two delta formations but the dots cannot be considered as obstructions crossing the line of flow at right angles. This precludes the classification of the central pocket loop type of whorl.

Figure 324 is a loop, the two recurving ridges have appendages and are considered spoiled. The pattern cannot, therefore, be a whorl even though two delta formations are present.

[Illustration: 323]

[Illustration: 324]

Figure 325 is classified as a tented arch. If examined closely the pattern will be seen to have an appendage abutting at a right angle between the shoulders of each possible recurve. Thus no sufficient recurve is present.

Figure 326 is a plain arch. There is present no angle which approaches a right angle. Points A, B, and X are merely bifurcations rather than an abutment of two ridges at an angle.

[Illustration: 325]

[Illustration: 326]

Figure 327 is a tented arch, not because of the dot, however, as it cannot be considered an upthrust. The tented arch is formed by the angle made when the curving ridge above the dot abuts upon the ridge immediately under and to the left of the dot.

[Illustration: 327]

CHAPTER III 58

[Illustration: 328]

Figure 328 consists of two separate looping ridge formations in juxtaposition upon the same side of a common delta. This pattern cannot be called a double loop as there is no second delta formation. In order to locate the core, the two looping ridges should be treated as one loop with two rods in the center. The core is thus placed on the far rod (actually on the left shoulder of the far loop), resulting in a ridge count of four (fig. 49).

[Illustration: 329]

[Illustration: 330]

Figure 329 is a loop of three counts. It cannot be classified as a whorl as the only recurve is spoiled by the appendage abutting upon it at the point of contact with the line of flow.

Figure 330 is a plain arch as there is no upthrust (an upthrust must be an ending ridge), no backward looping turn, and no two ridges abutting upon each other at a sufficient angle.

Figure 331 is a plain arch. The ending ridge at the center does not rise at a sufficient angle to be considered an upthrust, and it does not quite meet the ridge toward which it is flowing and therefore forms no angle.

Figure 332 is a plain arch. There are two ending ridges, but no separate delta formation is present.

[Illustration: 331]

[Illustration: 332]

[Illustration: 333]

[Illustration: 334]

Figure 333 is a plain arch. The rising ridge at the center is curved at the top forming no angle, and does not constitute an upthrust because it is not an ending ridge.

Figure 334 is a whorl of the double loop type. Two loops and two deltas are present. It is unusual because the loops are juxtaposed instead of one flowing over the other, and one delta is almost directly over the other. The tracing is a meeting tracing.

Figure 335 is a tented arch. Although there is a looping ridge, no ridge count can be obtained. The core is placed upon the end of the ridge abutting upon the inside of the loop, and so the imaginary line crosses no looping ridge, which is necessary.

Figure 336 is a plain arch. The ending ridge at the center cannot be considered an upthrust because it does not deviate from the general direction of flow of the ridges on either side. No angle is present as the ending ridge does not abut upon the curving ridge which envelopes it.

[Illustration: 335]

[Illustration: 336]

[Illustration: 337]

[Illustration: 338]

[Illustration: 339]

Figure 337 is a plain arch because the dot cannot be considered a delta as it is not as thick and heavy as the surrounding ridges.

Figure 338 is a tented arch consisting of two ending ridges and a delta. The short ending ridge is considered a ridge because it is slightly elongated and not a mere dot.

In figure 339, the only question involved is where to stop tracing. The rule is: *when tracing on a ridge with an upward trend, stop at the point on the upward trend which is nearest to the right delta.* X is the point in this pattern.

In figure 340, the question involved is also one of tracing. In this pattern, the tracing is not on a ridge with an upward trend. The tracing, therefore, is continued until a point nearest to the right delta, or the right delta itself, is reached. This tracing is a meeting tracing.

[Illustration: 340]

There are a few constantly recurring patterns which, though not questionable or doubtful as they appear, present a peculiarly difficult problem in classifying. The patterns referred to are usually double loops, though accidental whorls and loops sometimes present the same problems. The difficulty arises when a loop is so elongated that the recurve does not appear until near the edge of a fully rolled impression or an impression that is rolled unusually far, as in figures 341 to 344.

[Illustration: 341]

[Illustration: 342]

[Illustration: 343]

[Illustration: 344]

Figure 341, if classified as it appears, would be an accidental whorl. Figures 342 and 343 would be double loops, and illustration 344, a loop. It will be observed that these prints are rolled more fully than normal. If, however, the next time the prints are taken, they are not rolled quite so far, the patterns would require a very different classification, and would show no indication of any need for referencing to their true classification. The result would be a failure to establish an identification with the original prints. The only way in which such an error may be avoided is to classify such impressions as they would appear if not so fully rolled, and to conduct a reference search in the classification which would be given to the prints when rolled to the fullest extent. Applying this rule, illustration 341 is a tented arch, referenced to a whorl. Figures 342 and 343 are loops, referenced to whorls. Figure 344 is a plain arch, referenced to a loop.

CHAPTER III

No set rule can possibly be devised to enable a classifier to know with certainty where to draw the line when it is doubtful which classification should be given such a print. Individual judgment is the only standard. The test is: *if the pattern, in the opinion of the classifier, is rolled to only a normal width, it should be classified as it appears. If it seems to be rolled to a width beyond the normal degree, it should be classified as if rolled only to the normal degree.* Age, weight, size of fingers (as seen in the plain impressions), heaviness of the ridges, and experience of the technician in taking fingerprints are all factors in arriving at the correct conclusion. The necessity for exercising the utmost care in dealing with this type of pattern cannot be too highly emphasized.

[Illustration: 345]

[Illustration: 346]

The patterns in figures 345 and 346 also have a second loop near the edge of the impression. In these two patterns, however, the second loop is very near the delta and consequently will almost invariably appear even though not rolled to the fullest extent. The foregoing rule is not applied to this type of impression. Both are classified as a whorl and referenced to a loop to take care of the rare contingency of nonappearance.

CHAPTER IV

The Classification Formula and Extensions

The classification formula

At this point it is necessary to mention that when prints are classified, markings are indicated at the bottom of each finger block to reflect the type. The following symbols are used:

- Under the index fingers the appropriate capital letters should be placed for every pattern except the ulnar loop.

- Under all other fingers, the appropriate small letter should be placed for every pattern except the ulnar loop and the whorl as follows:

Arch a Tented Arch t Radial Loop r

- Ulnar loops in any finger are designated by a diagonal line slanting in the direction of the loop.

- Whorls in any finger are designated by the letter "W". The classification formula may be composed of the following divisions:

1. Primary 2. Secondary 3. Subsecondary 4. Major 5. Final 6. Key

The positions in the classification line for these divisions when completely applied are as illustrated:

Key Major Primary Secondary Subsecondary Final
-- Divisions
Classification Classification Classification

20 M 1 U IOI 10
-- L 1 U IOI

Second subsecondary classification Key Major Primary Secondary Subsecondary Final
-- Divisions

CHAPTER IV

Classification Classification Classification

SLM --- MMS 4 O 5 U IOI 10
-- I 17 U IOI

THE PRIMARY CLASSIFICATION: For the purpose of obtaining the primary classification, numerical values are assigned to each of the ten finger spaces as shown in figure 347. Wherever a whorl appears it assumes the value of the space in which it is found. Spaces in which types of patterns other than whorls are present are disregarded in computing the primary.

The values are assigned as follows:

Fingers No. 1 and No. 2   16

Fingers No. 3 and No. 4   8

Fingers No. 5 and No. 6   4

Fingers No. 7 and No. 8   2

Fingers No. 9 and No. 10   1

[Illustration: 347]

LEAVE THIS SPACE BLANK| |SEX +---------------+ | |_____|FBI No. | | |RACE +--------------+ |LAST NAME FIRST NAME MIDDLE NAME| ---------------------------|----------------------------------+----------
SIGNATURE OF PERSON |CONTRIBUTOR |ALIASES |HT. |WT. FINGERPRINTED |AND ADDRESS | |(IN.)| | |___|__ | | |DATE OF ----------------------------| | |BIRTH RESIDENCE OF PERSON | | |_____FINGERPRINTED | | |HAIR |EYES | | | |
----------------------------|---
OCCUPATION |ARREST NUMBER|LEAVE THIS SPACE BLANK | | -----------------------------|-------------| SCARS AND MARKS |PLACE OF |
|BIRTH | 29
|-------------|CLASS_____ ----------------------------|CITIZEI
| 19 SIGNATURE OF OFFICIAL |DATE | | TAKING FINGERPRINTS |

CHAPTER IV 64

|||CHECK IF | | | NO CRIMINAL|REF._____ | |
RECORD IS | | | DESIRED |
-- 1. RIGHT
THUMB|2. RIGHT INDEX|3. RIGHT |4. RIGHT RING |5. RIGHT
LITTLE | | MIDDLE | | |N 16| |N 8|
[Illustration]|[Illustration]|[Illustration]|[Illustration]|[Illustration] D 16| |D
8| |D 4 -- W | W
| \ | W | \ -- 1.
LEFT THUMB |2. LEFT INDEX |3. LEFT |4. LEFT RING |5. LEFT
LITTLE N 4| | MIDDLE | | | |N 2| |N 1
[Illustration]|[Illustration]|[Illustration]|[Illustration]|[Illustration] |D 2| |D 1|
-- W | W | / | / | /

In figure 347, it will be observed that the odd fingers (Nos. 1, 3, 5, 7, 9) contain the letter D, and the even fingers (Nos. 2, 4, 6, 8, 10) contain the letter N. The D indicates that the values of these fingers relate to the denominator, the N that they relate to the numerator. The summation of the numerical values of the whorl type patterns, if any, appearing in fingers 1, 3, 5, 7, 9, plus one, is the denominator of the primary. The summation of the values of the whorls, if any, in fingers 2, 4, 6, 8, 10, plus one, is the numerator of the primary. Where no whorl appears in a set of impressions, the primary, therefore, would be 1 over 1. The 1 that is assigned to the numerator and the denominator when no whorls appear is also added, for consistency, to the value of the whorls when they do appear. It will be understood why it was originally assigned to the no-whorl group when it is considered how easily a zero might be confused with an O, which is the symbol used for an outer whorl tracing.

To obtain the primary for the prints in figure 347, the number of whorls appearing in the odd fingers is ascertained to be 2. Their positions are noted (1 in No. 1 and 1 in No. 7) and the values assigned to whorls appearing in those fingers are added together (16 plus 2 = 18). To this sum the arbitrary 1 is added, giving us the total of 19, which constitutes the denominator for this set of prints. To get the numerator, it is ascertained that there are 3 whorls appearing in the even fingers (2, 4 and 6), the values of which are added together (16 plus 8 plus 4 = 28). To this sum the 1 is added, giving a numerator of 29, and a complete

CHAPTER IV

primary of 29 over 19.

By the word "whorl" is meant all types of whorls, including plain whorls, central pocket loops, double loops and accidentals. The tracing of the whorl does not enter into the determination of the primary.

The method of obtaining the primary can probably be shown best by illustrations. For example, assume that there is a whorl in the right index finger only. The value of a whorl in this finger is 16. When 1 over 1 is added the resulting primary is 17 over 1. If a whorl appears in the right thumb and right index finger, the value is 16 over 16 plus 1 over 1 giving a primary of 17 over 17. If whorls appear in both index fingers, the value is 16 over 2 plus 1 over 1 giving a primary of 17 over 3. When whorls appear in both thumbs and both index fingers, the primary is 21 over 19 and is obtained by the addition 16 plus 4 plus 1 over 16 plus 2 plus 1. If whorls appear in all 10 fingers, the primary is 32 over 32 (16 plus 8 plus 4 plus 2 plus 1 plus 1 over 16 plus 8 plus 4 plus 2 plus 1 plus 1). It will be noted that the primary classifications extend from 1 over 1 in the no-whorl group to 32 over 32 in the all-whorl group, providing 1,024 possible combinations. This does not mean that there are 1,024 even subdivisions of prints according to these primaries. Just as there is a preponderance of loops when the types of patterns are considered, there is also a preponderance of certain primaries, notably: the 1 over 1 primary, or no-whorl group; the 17 denominator; the 19 denominator; the 28 denominator, of which the 31 over 28 group is the largest; and the 32 denominator, including 2 large primary groups namely, 31 over 32 and 32 over 32. As a matter of fact, the 1 over 1 group, as a whole, contains over 25 percent of the total number of prints filed in the FBI. On the other hand, there are a number of primaries which rarely appear. It follows, therefore, that when a print is classified in one of these larger groups it is necessary to complete the classification to a greater extent than is necessary in the more unusual primaries, so that the group to be searched is small enough for convenience.

In connection with the counting of whorl values to obtain the primary, it might be noted that when the whorls outnumber the other patterns more speed can be achieved by counting those patterns and subtracting rather than by adding the whorls. This procedure should not be followed until enough experience is acquired so that it may be noted at a glance where whorls are not present.

The experienced classifier can tell in what fingers whorls are present by a glance at a primary classification. For example, a primary of 5 over 17 could mean that there are whorls in the thumbs only.

[Illustration: 348]

LEAVE THIS SPACE BLANK| |SEX +---------------+ | |_____|FBI No. | | |RACE +---------------+ |LAST NAME FIRST NAME MIDDLE NAME| ----------------------------|---------------------------------+---------- SIGNATURE OF PERSON |CONTRIBUTOR |ALIASES |HT. |WT. FINGERPRINTED |AND ADDRESS | |(IN.)| | |___|__ | | |DATE OF ----------------------------| | |BIRTH RESIDENCE OF PERSON | | |_____FINGERPRINTED | | |HAIR |EYES | | | | -----------------------------|---
OCCUPATION |ARREST NUMBER|LEAVE THIS SPACE BLANK | | -----------------------------|-------------| SCARS AND MARKS |PLACE OF | |BIRTH | 9 R |--------------|CLASS_____ ----------------------------|CIT. | 2 R SIGNATURE OF OFFICIAL |DATE | | TAKING FINGERPRINTS | |||CHECK IF | | | NO CRIMINAL|REF._____ | | RECORD IS | | | DESIRED |
--- 1. RIGHT THUMB|2. RIGHT INDEX|3. RIGHT |4. RIGHT RING |5. RIGHT LITTLE | | MIDDLE | | | | | | [Illustration]|[Illustration]|[Illustration]|[Illustration]|[Illustration] | | | | --- \| R |\| W |\ --- 1. LEFT THUMB |2. LEFT INDEX |3. LEFT M|4. LEFT RING |5. LEFT LITTLE 18| 10| MIDDLE I| I| 13 | | | | [Illustration]|[Illustration]|[Illustration]|[Illustration]|[Illustration] | | | | --- / | R | / | W | / ---

THE SECONDARY CLASSIFICATION: After the primary classification, the fingerprints are subdivided further by using a secondary classification. Before going into detail, it should be noted that after the

CHAPTER IV

primary is obtained the entire remaining portion of the classification formula is based upon the arrangement of the impressions appearing in the right hand as the numerator over the impressions appearing in the left hand as the denominator. The arrangement of the even over the uneven fingers is discarded after the primary is obtained. The secondary classification appears just to the right of the fractional numerals which represent the primary. It is shown in the formula by capital letters representing the basic types of patterns appearing in the index fingers of each hand, that of the right hand being the numerator and that of the left hand being the denominator (fig. 348). There are five basic types of patterns which can appear.

1. Arch A 2. Tented Arch T 3. Radial Loop R 4. Ulnar Loop U 5. Whorl W

[Illustration: 349]

LEAVE THIS SPACE BLANK| |SEX +---------------+ | |_____|FBI No. | | |RACE +--------------+ |LAST NAME FIRST NAME MIDDLE NAME| ----------------------------|----------------------------------+--------- SIGNATURE OF PERSON |CONTRIBUTOR |ALIASES |HT. |WT. FINGERPRINTED |AND ADDRESS | |(IN.)| | | |___|__ | | |DATE OF ---------------------------| | |BIRTH RESIDENCE OF PERSON | | |_____FINGERPRINTED | | |HAIR |EYES | | | | ---------------------------|------------------------------------- OCCUPATION |ARREST NUMBER|LEAVE THIS SPACE BLANK | | ----------------------------|-------------| SCARS AND MARKS |PLACE OF | |BIRTH | 1 R |-------------|CLASS_____----------------------------|CITIZEN | 1 aU SIGNATURE OF OFFICIAL |DATE | | TAKING FINGERPRINTS | |||CHECK IF | | | NO CRIMINAL|REF._____ | | RECORD IS | | | DESIRED | -- 1. RIGHT THUMB|2. RIGHT INDEX|3. RIGHT |4. RIGHT RING |5. RIGHT LITTLE | | MIDDLE | | | | | [Illustration]|[Illustration]|[Illustration]|[Illustration]|[Illustration] | | | | -- \ | R | \ | \ | \ -- 1. LEFT THUMB |2. LEFT INDEX |3. LEFT M|4. LEFT RING |5. LEFT LITTLE

CHAPTER IV

18/ 10/ MIDDLE I/ I/ 13 / / / /

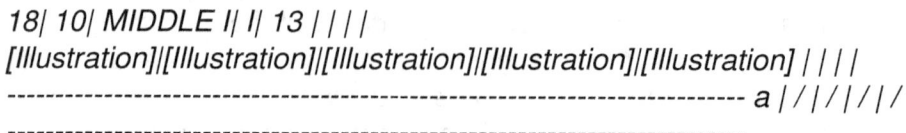

[Illustration]/[Illustration]/[Illustration]/[Illustration]/[Illustration] / / / /
-- *a / / / / / /*

SECONDARY CLASSIFICATION (SMALL-LETTER GROUP): Prints with an arch or tented arch in any finger or a radial loop in any except the index fingers constitute the small-letter group of the secondary classification. Such "small letters," with the exception of those appearing in the index fingers, are brought up into the classification formula in their proper relative positions immediately adjacent to the index fingers (fig. 349). A dash is used to indicate the absence of each small letter between the index fingers and another small letter or between two small letters, as

1 aUa-t 1 aU-t. ------- and ------ 1 R-a 1 U-a

Thus, if a radial loop appears in the right thumb, the small letter "r" would be brought up in the numerator column of the classification formula and placed just to the left of the capital letter representing the index finger. Similarly, if an arch or tented arch or a radial loop would appear in the middle, ring, or little finger of the hand, the small letter representing such a pattern would be placed on the classification line to the right of the secondary in the numerator column if the letter is present in the right hand, and in the denominator column if in the left hand. When two or more small letters of the same type occur immediately adjacent to each other, they are indicated thus:

1 rU-2a 1 aTa-a. ------- and ------- 1 tU3a 1 tA2at

The small-letter groups are of vital importance to the classification system, as they are of relatively infrequent occurrence, constituting approximately 7 to 10 percent of all patterns. Generally speaking, since these patterns are of such rare occurrence, their very presence often enables the classifier to dispense with the usual subsecondary classification and the major division which in the majority of cases are used in the larger groups.

CHAPTER IV

THE SUBSECONDARY CLASSIFICATION (GROUPING OF LOOPS AND WHORLS): In classifying prints it is necessary to subdivide the secondary groups. This is accomplished by grouping according to the ridge counts of loops and the ridge tracings of whorls. The first of the groups filed in order, which it will be necessary to so subdivide, would ordinarily be the

1 R --- 1 R

group where no small letters appear. The Federal Bureau of Investigation, however, has found it necessary to extend this division to many of the small-letter groups which become cumbersome. The subsecondary is placed on the classification line just to the right of the secondary. Ridge counts are translated into small and large, represented by symbols I and O. The whorl tracings are brought up as I, M, or O denoting inner, meeting or outer ridge tracings of the whorl types. Only six fingers may be involved in the subsecondary--numbers 2, 3, 4, 7, 8, and 9.

A ridge count of 1 to 9, inclusive, in the index fingers is brought up into the subsecondary formula as I. A count of 10 or more is brought up as O. In the middle fingers a count of from 1 to 10, inclusive, is brought up as I, and 11 or more is O. In the ring fingers a count of from 1 to 13 is brought up as I, and 14 or more is O. A loop subsecondary could appear in the classification formula as

OIO. --- IIO

Analyzing this example of a subsecondary, one will know that in the index, middle, and ring fingers of the right hand there are counts of over 9, under 11, and over 13, while in the left hand there are in the index, middle, and ring fingers, counts of under 10, under 11, over 13, respectively. The subsecondary classification, therefore, relates to the groupings of the prints, and no difficulty should be experienced in ascertaining whether the I and O arrangement in the subsecondary relates to loops or whorls when analyzing a classification, because this information can be obtained from the primary classification. Figure 350 is an example illustrating the subsecondary in addition to other divisions of the classification formula.

CHAPTER IV

[Illustration: 350]

LEAVE THIS SPACE BLANK| |SEX +--------------+ | |_____|FBI No. | | |RACE +--------------+ |LAST NAME FIRST NAME MIDDLE NAME| ----------------------------|--------------------------------+---------- SIGNATURE OF PERSON |CONTRIBUTOR |ALIASES |HT. |WT. FINGERPRINTED |AND ADDRESS | |(IN.)| | | |___|__ | | |DATE OF ---------------------------| | |BIRTH RESIDENCE OF PERSON | | |_____FINGERPRINTED | | |HAIR |EYES | | | | ---------------------------|------------------------------------- OCCUPATION |ARREST NUMBER|LEAVE THIS SPACE BLANK | | ------------------------------|-------------| SCARS AND MARKS |PLACE OF | |BIRTH | 26 5 R OOO 12 |-------------|CLASS_____ --------------------------|CITIZEN | 12 W MOI SIGNATURE OF OFFICIAL |DATE | | TAKING FINGERPRINTS | |||CHECK IF | | | NO CRIMINAL|REF._____ | | RECORD IS | | | DESIRED | -- 1. RIGHT THUMB|2. RIGHT INDEX|3. RIGHT |4. RIGHT RING |5. RIGHT LITTLE | | MIDDLE | | 26| 12| 0| 17| 12 [Illustration]|[Illustration]|[Illustration]|[Illustration]|[Illustration] --- \ | R | W | \ | \ --- 1. LEFT THUMB |2. LEFT INDEX |3. LEFT |4. LEFT RING |5. LEFT LITTLE | | MIDDLE | | I| M | 18| I| 15 [Illustration]|[Illustration]|[Illustration]|[Illustration]|[Illustration] --- W | W | / | W | / ---

The chart, figure 351, will illustrate the manner in which the ridge counts are translated into the symbols I and O so they may be grouped and sequenced with the whorl tracings I, M and O.

THE MAJOR DIVISION: *The major division is placed just to the left of the primary in the classification formula. Where whorls appear in the thumbs the major division reflects the whorl tracings just as the subsecondary does. For example, a major division of I over M in the primary 5 over 17 would reflect an inner-traced whorl over a meeting-traced whorl in the thumbs. Where loops appear in the* thumbs, however, a table is used to translate the ridge counts into the small, medium, or large groups, designated by the letters S, M, L. An expanding table is used for the right thumb when large-count loops appear in the left thumb, as shown in the chart (fig. 351). This table is used because it affords a more equitable distribution of prints as a whole, for filing purposes within the groups indicated.

[Illustration: 351. Classification Chart]

--RIGHT HAND--
--- R THUMB |R INDEX |R MIDDLE |R RING |R LITTLE
--- WHEN LEFT THUMB| 1-9 = I | 1-10 = I | 1-13 = I | IS 16 OR LESS |10 AND OVER = O|11 AND OVER = O|14 AND OVER = O| 1-11 = S | | | | 12-16 = M | | | | 17 AND OVER = L| | | | --------------------RIDGE COUNT OF SECOND SUBSECONDARY--------------------- WHEN LEFT THUMB| 1-5 = S | 1-8 = S | 1-10 = S | IS 17 OR OVER | 6-12 = M | 9-14 = M |11-18 = M | 1-17 = S |13 AND OVER = L|15 AND OVER = L|19 AND OVER = L| 19-22 = M | | | | 23 AND OVER = L| | | |

--LEFT HAND--
--- L THUMB |L INDEX |L MIDDLE |L RING |L LITTLE
--- 1-11 = S | | | | 12-16 = M |<------------------VALUES SAME AS ABOVE------------------> 17 AND OVER = L| | | |

Table for major divisions of loops:

Left thumb denominator Right thumb numerator

{ 1 to 11, inclusive, S (small). 1 to 11, inclusive, S (small) { 12 to 16, inclusive, M (medium). { 17 or more ridges, L (large).

{ 1 to 11, inclusive, S (small). 12 to 16, inclusive, M (medium) { 12 to 16, inclusive, M (medium). { 17 or more ridges, L (large).

{ 1 to 17, inclusive, S (small). 17 or more ridges, L (large) { 18 to 22, inclusive, M (medium). { 23 or more ridges, L (large).

The fingerprint card appearing in figure 352 shows a major division of L over L, which is obtained by counting the ridges (24 in the right thumb and 18 in the left thumb) which, according to the table, is translated into L in both thumbs.

THE FINAL: It is, of course, desirable to have a definite sequence or order of filing the prints within the subdivided groups. This order is attained through the use of the final, which is based upon the ridge count of the loop in the right little finger. It is indicated at the extreme right of the numerator in the classification. Note figure 352. If a loop does not appear in the right little finger, a loop in the left little finger may be used. It is then indicated at the extreme right of the denominator (fig. 353). If no loops appear in the little fingers, a whorl may be used to obtain a final, counting from left delta to core if in the right hand and from right delta to core if in the left hand. If there are two or more cores (usually applies to accidental whorls), the ridge count is made from left delta (right hand) or right delta (left hand) to the core which is the least number of ridges distant from that delta. An exception is made in the case of the double loop. The double loop is counted from the delta to the core of the upright loop. Where loops of a double loop are horizontal, the nearest core is used. Should both little fingers be a or t, no final is used. The use of a whorl in a little finger for a final is required only in connection with a large group or collection of prints, such as the 32 over 32 primary.

[Illustration: 352]

LEAVE THIS SPACE BLANK| |SEX +---------------+ | |_____|FBI No. | | |RACE +--------------+ |LAST NAME FIRST NAME MIDDLE NAME| -----------------------------|----------------------------------+---------- SIGNATURE OF PERSON |CONTRIBUTOR |ALIASES |HT. |WT. FINGERPRINTED |AND ADDRESS | |(IN.)| | | |___|__ | | |DATE OF ---------------------------| | |BIRTH RESIDENCE OF PERSON | | |_____FINGERPRINTED | | |HAIR |EYES | | | | ---------------------------|--- OCCUPATION |ARREST NUMBER|LEAVE THIS SPACE BLANK | |

CHAPTER IV

```
LLL ---------------------------/-------------/ LMM  SCARS AND MARKS
/PLACE OF / /BIRTH / 24 L I R O O O 17
/-------------/CLASS_____  ---------------------------/CITIZEN
/ L I R O O O SIGNATURE OF OFFICIAL /DATE / / TAKING
FINGERPRINTS / ///CHECK IF / / / NO
CRIMINAL/REF._____ / / RECORD IS / / /
DESIRED / ---------------------------------------------------------------------- 1.
RIGHT 24/2. RIGHT 13/3. RIGHT 31/4. RIGHT 21/5. RIGHT 17
THUMB / INDEX / MIDDLE / RING / LITTLE / / / /
[Illustration]/[Illustration]/[Illustration]/[Illustration]/[Illustration]
------------------------------------------------------------------------ \ / R / \ / \ / \
-------------------------------------------------------------- 1. LEFT 18/2.
LEFT 16/3. LEFT 13/4. LEFT 18/5. LEFT 20 THUMB / INDEX /
MIDDLE / RING / LITTLE / / / /
[Illustration]/[Illustration]/[Illustration]/[Illustration]/[Illustration]
---------------------------------------------------------------------- / / R / / / / /
------------------------------------------------------------------
```

THE KEY: The key is obtained by counting the ridges of the first loop appearing on the fingerprint card (beginning with the right thumb), exclusive of the little fingers which are never considered for the key as they are reserved for the final. The key, no matter where found, is always placed to the extreme left of the numerator of the classification formula (fig. 353).

Extensions

THE SECOND SUBSECONDARY CLASSIFICATION: When a group of fingerprints becomes so large that it is cumbersome and unwieldy, even though fully extended, it can be subdivided further by using a second subsecondary division, which is brought up into the classification formula directly above the subsecondary, and for which the symbols S, M and L are used. The following table is used:

Index Middle Ring

1 to 5, inclusive, S. 1 to 8, inclusive, S. 1 to 10, inclusive, S. 6 to 12, inclusive, M. 9 to 14, inclusive, M. 11 to 18, inclusive, M. 13 or more, L. 15 or more, L. 19 or more, L.

CHAPTER IV

If this table is referred to, a study of figure 352 will demonstrate the use of the second subsecondary.

[Illustration: 353]

```
LEAVE THIS SPACE BLANK| |SEX +--------------+ | |_____|FBI No.
| | |RACE +--------------+ |LAST NAME FIRST NAME MIDDLE NAME|
---------------------------|---------------------------------+---------- SIGNATURE
OF PERSON |CONTRIBUTOR |ALIASES |HT. |WT. FINGERPRINTED
|AND ADDRESS | |(IN.)| | |____|__ | | |DATE OF
---------------------------| | |BIRTH RESIDENCE OF PERSON | |
|_____FINGERPRINTED | | |HAIR |EYES | | | |
---------------------------|--------------------------------------
OCCUPATION |ARREST NUMBER|LEAVE THIS SPACE BLANK | |
---------------------------|--------------| SCARS AND MARKS |PLACE OF |
|BIRTH | 22 M 11 U OOO
|-------------|CLASS_____ ----------------------------|CITIZEN
| L 6 U OMI 13 SIGNATURE OF OFFICIAL |DATE | | TAKING
FINGERPRINTS | |||CHECK IF | | | NO
CRIMINAL|REF._____ | | RECORD IS | | |
DESIRED | ----------------------------------------------------------------------- 1.
RIGHT THUMB|2. RIGHT INDEX|3. RIGHT |4. RIGHT RING |5.
RIGHT LITTLE 22| 11| MIDDLE 19| O| O | | | |
[Illustration]|[Illustration]|[Illustration]|[Illustration]|[Illustration] | | | |
----------------------------------------------------------------------- \ | \ | \ | W | W
----------------------------------------------------------------------- 1. LEFT
THUMB |2. LEFT INDEX |3. LEFT M|4. LEFT RING |5. LEFT LITTLE
18| 10| MIDDLE I| I| 13 | | | |
[Illustration]|[Illustration]|[Illustration]|[Illustration]|[Illustration] | | | |
----------------------------------------------------------------------- / | / | W | W | /
-----------------------------------------------------------------------
```

WCDX EXTENSION: In the extension used in the Federal Bureau of Investigation for the large whorl groups, the type of whorl is designated by the symbols W, C, D, or X for the index fingers and w, c, d, or x for all other fingers, according to its classification as defined in figure 354. These symbols are used for subclassification purposes only and are brought up into the classification formula directly above the subsecondary in their respective positions, the right hand being the

CHAPTER IV

numerator, the left hand being the denominator.

SPECIAL LOOP EXTENSION: In the all-loop group

(1R-U) ------ (1R-U),

the following special loop extension may be used, utilizing the ridge counts in fingers Nos. 2, 3, 4, 7, 8, 9, and, if necessary, No. 10:

Ridge Counts Value

1 to 4, inclusive 1 5 to 8, inclusive 2 9 to 12, inclusive 3 13 to 16, inclusive 4 17 to 20, inclusive 5 21 to 24, inclusive 6 25 and over 7

The resulting values in this extension are brought up into the classification formula directly above the subsecondary in their respective positions, the right hand being the numerator, the left hand being the denominator.

In addition to the extensions already mentioned, fingerprint groups may be divided into male and female, and by age (either by year or by arbitrarily setting an age limit, beyond which a print bearing such an age would be filed separately in a "Reference" or a "Presumptive Dead" file).

In the files of the Federal Bureau of Investigation, all prints bearing an age of 55 through 74 are filed in the "Reference" group and all prints bearing an age of 75 years or more are filed in the "Presumptive Dead" file. Persons 75 years of age or older, in regard to crime, may be considered as generally inactive and thus are filed as "Presumptive Dead." Such a group provides for removing from the other files the cards concerning those of whom no notice is ever received as to death.

A separate file should be maintained for deceased persons, for possible future reference.

A separate file should be maintained for all prints bearing amputations and which have an unequivocal statement or marking from the *contributor to that effect.*

Permanent scars also may be utilized for this purpose, giving three more groupings: those prints having permanent scars in the right hand, those having a scar in the left, and those in which scars appear in both hands. A separate file may be maintained for mutilated prints whether or not the permanent-scar division is used. This is usually composed of prints so badly mutilated, or so mutilated about the cores and deltas, that intentional mutilation is suspected.

[Illustration: 354]

```
+--------------+ +--------------+ +--------------+ +--------------+
|CENTRAL POCKET| | DUAL LOOP    | | ACCIDENTAL   | | -W-          | | WHORL |
| -D-          | | -X-          | |              | |              | | LOOP  |
| [Illustration]| | [Illustration]| | [Illustration]| | [Illustration]|
+--------------+ +--------------+ +--------------+ +--------------+
PATTERN HAVING   PATTERN HAVING   TWO SEPARATE    TWO OR
MORE ONE CORE.   LINE ONE CORE.   LINE AND        DISTINCT
DIFFERENT TYPES  DRAWN FROM       DRAWN FROM      LOOPS IN ONE
PATTERN. DELTA TO DELTA  DELTA TO DELTA  ANY UNUSUAL
CUTS ONE OR      CUTS NO PATTERN  NOT MORE        RECURVES
RECURVES         DEFINED          IN OTHER CLASSIFICATIONS
```

CHAPTER V

Classification of Scarred Patterns--Amputations--Missing at Birth

Classification of scarred patterns

Emphasis should be placed upon the necessity for fully referencing all scarred patterns. In connection with their proper classification, the following rules should be observed:

- When an impression is so scarred that neither the general type of pattern nor the ridge tracing or count can be determined with reasonable accuracy, the impression should be given both the general type value and the subclassification value of the corresponding finger of the other hand.

- When an impression is partially scarred, i.e., large scars about the core so that the general type cannot be determined with reasonable accuracy, but the ridges allow reasonably accurate subclassifications by ridge tracings or counting, the impression should be given the primary value of the pattern of the corresponding finger and the subclassification value as indicated by the ridges of partially scarred impressions.

- When an impression is partially scarred and the general type of pattern can be determined with reasonable accuracy, but the ridges cannot be traced or counted so as to fall within the proper subsecondary classification, the impression should be given the ridge count or tracing value of the corresponding finger of the other hand, if the corresponding finger is of the same general type. If the corresponding finger is not of the same general type, the scarred impression should be given the probable value and referenced to all other possibilities.

- When an impression is so scarred that neither the general type of pattern nor the ridge tracing or count can be determined with reasonable accuracy, and it so happens that the corresponding finger of the other hand is similarly scarred, both patterns are given the arbitrary value of whorls with meeting tracings.

CHAPTER V

In figure 355, the pattern is entirely obliterated. It could have been a small whorl, a small ulnar or radial loop, an arch, or a tented arch. If the opposite finger were an arch or tented arch or whorl, this impression would be classified as arch, tented arch, or whorl (with the same tracing). If the opposite finger were a small-count loop, this would be classified as a loop of the same count. If the opposite finger were a large-count loop, this impression would be given the count of the opposite finger even though it could never have had that count. If the opposite finger were scarred in the same fashion or were amputated or missing, both impressions would be classified as whorls with meeting tracings.

In figure 356, the general type of the pattern could have been loop (ulnar if in the right hand) or whorl. If the opposite finger were a whorl this would be classified as a whorl, and with the same tracing. If a radial loop were opposite, this would be classified as an ulnar loop (if in the right hand). The ridge count can be obtained with a fair degree of accuracy. If an arch or tented arch were opposite, this impression would be classified as a loop because it looks as if it had been a loop.

[Illustration: 355]

[Illustration: 356]

[Illustration: 357]

[Illustration: 358]

In figure 357, the ridge count cannot be determined accurately but it would be classified as a loop, no matter what the opposite finger might be. If the opposite finger were a loop with a count of from 6 to 17, this impression would be given that count. If the count of the opposite loop were less or more than 6 to 17, the count for this finger would be given I or O in the subsecondary classification depending upon whether the opposite finger was I or O, but would not be given less than 6 nor more than 17 counts as its possibility is limited to those counts.

A pattern with a scar similar to either scar in figures 358 and 359 would always be given a loop as it could be seen readily that there was no *possibility of its having been any other type of pattern.*

[Illustration: 359]

Classification of amputations and fingers missing at birth

When one or more amputations appear upon a fingerprint card, it may be filed separately from those having no amputations in order to facilitate searching. It is to be noted that before it may be filed in the amputation group, the card must contain a definite and unequivocal statement or marking by the contributor to the effect that a certain finger or fingers have been amputated or were missing at birth. This prevents the appearance on later cards of impressions of fingers thought to have been amputated but which in reality were merely injured and bandaged when previous prints were submitted.

If one finger is amputated, it is given a classification identical with that of the opposite finger, including pattern and ridge count, or tracing, and referenced to every other possible classification.

If two or more fingers are amputated, they are given classifications identical with the fingers opposite, with no additional references.

If two amputated fingers are opposite each other, both are given the classification of whorls with meeting tracings.

When a fingerprint card bearing a notation of fingers missing at birth is classified, the missing fingers should be treated as amputations in that they are given the identical classifications of the opposite fingers and are filed in the amputation group. As these fingers are missing from a prenatal cause, they would have always received the identical classification of the opposite finger on any previous occasion.

If all 10 fingers are amputated or missing at birth, the classification will be

M 32 W MMM. ----------- M 32 W MMM

If both hands are amputated or missing at birth, the footprints should be taken as they, too, bear friction ridges with definite patterns. A footprint file is maintained by the FBI for identification purposes in instances where the subject has all fingers amputated or missing at birth.

Partially amputated fingers often present very complex problems and careful consideration should be given to them. The question often arises as to the appropriate groups in which they should be filed, i.e., amputations or nonamputations. As no definite rule may be applied, it is a matter of experience and judgment as to their preferred classification.

In those instances in which a partially amputated finger has half or more than half of the pattern area missing, it is given the classification of the opposite finger. It will be filed in the amputation group under the classification of the opposite finger and reference searches should be conducted in all possible classifications in the nonamputation groups. If two or more of the fingers are amputated in this manner, they are given the classification of the opposite fingers only and are governed by the rules concerning amputations.

Generally, a "tip amputation," or one which has less than half of the first joint amputated, will always be printed in the future. Therefore, a partially amputated finger with less than half of the pattern area missing is classified as it appears and is referenced to the opposite finger. It will be filed in the nonamputation group and reference searches should be conducted under the classification of the opposite finger, and in the amputation group. It must be referenced this way even though it never could have originally had the classification of the opposite finger.

Classification of bandaged or imprinted fingers

As noted in the chapter pertaining to "Problems in the Taking of Inked Fingerprints," an indication to the effect "recently injured, bandaged" is not sufficient to file a fingerprint card. It is obvious that a fingerprint card bearing these notations cannot be properly classified or filed. If the injury is temporary, and if possible, these prints should not be taken until after healing.

If fingers are injured to the extent that it is impossible to secure inked impressions by special inking devices, the unprinted fingers are given classifications identical with the classifications of the fingers opposite. If only one finger is lacking, reference searches should be conducted in every possible classification. If more than one finger is lacking, they should be given the classifications of the opposite fingers, but no reference searches should be conducted. If there are two lacking, opposite each other, they should be classified as whorls with meeting tracings.

If, however, in the case of an injured finger, observation is made of the ridges of the finger itself and indicated on the print, this classification should be, insofar as it is possible, utilized. For example, a missing impression labeled "ulnar loop of about 8 counts" by the individual taking the prints, should be searched in the subsecondary as both I and O but should not be referenced as a pattern other than a loop. If the finger is used as the final, or key, it should be searched enough counts on each side of 8 to allow for possible error in the counting by the contributor using his naked eye.

CHAPTER VI

Filing Sequence

The sequence must be arranged properly at all times to make possible the most accurate work. Prints are sequenced and filed in this order, according to:

I. Primary:

1 32. - to -- 1 32

In the primary classification the denominator remains constant until all numerator figures have been exhausted from 1 to 32. All prints with the primary 1 over 1 are filed together. These are followed by 2 over 1, 3 over 1, 4 over 1, etc., until 32 over 1 is reached. The next primary is 1 over 2, then 2 over 2, etc., until 32 over 2 is reached. Eventually, through the use of each denominator figure and the elimination of each numerator over each denominator, the 32 over 32 primary will be reached.

Even in the smaller collections of fingerprints, it will be found that the groups which are arranged under the individual primaries filed in sequence, from 1 over 1 to 32 over 32, will be too voluminous for expeditious searching.

II. Secondary:

A. Secondary small-letter group:

A rW3r. - to ---- A rW3r

Most intricate of all the individual sequences is the small-letter sequence. It is less difficult if the following method is used:

1. Sequence according to the patterns in the index fingers, grouped

A W. - to - A W

CHAPTER VI

When small letters are present, there are 25 possible combinations which can appear in the index fingers. They are as follows:

A T R U W - - - - - A A A A A

A T R U W - - - - - T T T T T

A T R U W - - - - - R R R R R

A T U R W - - - - - U U U U U

A T U R W - - - - - W W W W W

2. Within each group sequence:

a. The denominator, by--

(1) Count of the small letters (lesser preceding the greater).

(2) Position of the small letters (those to the left preceding those to the right).

(3) Type of small letter (sequence a, t, r).

b. The numerator, by--

(1) Count.

(2) Position.

(3) Type.

Thus

A T - precedes - A A

A A --- precedes --- rAt A3t

A A -- precedes -- aA Aa

CHAPTER VI 84

A A -- precedes -- At Ar

aA aAa --- precedes --- aAr aAr

rA Ar ---- precedes ---- aA2a aA2a

aAtat aAtar ----- precedes ----- tA3r tA3r

The following table represents the full sequence of the denominator of the group having A over A in the index fingers. The full sequence as listed may be used as the numerator for each denominator as set out below. Following the group with A over A in the index fingers is the group with T over A in the index fingers, the sequence being the same otherwise. Then R over A, U over A, A over T to rW3r over rW3r.

A tAra aA2at tA2tr aA tArt aA2ar tAtra tA tA2r aAata tAtrt rA rA2a
aAa2t tAt2r Aa rAat aAatr tAr2a At rAar aAara tArat Ar rAta aAart tArar
aAa rA2t aAa2r tArta aAt rAtr aAt2a tAr2t aAr rAra aAtat tArtr tAa rArt
aAtar tA2ra tAt rA2r aA2ta tA2rt tAr A3a aA3t tA3r rAa A2at aA2tr rA3a
rAt A2ar aAtra rA2at rAr Aata aAtrt rA2ar A2a Aa2t aAt2r rAata Aat
Aatr aAr2a rAa2t Aar Aara aArat rAatr Ata Aart aArar rAara A2t Aa2r
aArta rAart Atr At2a aAr2t rAa2r Ara Atat aArtr rAt2a Art Atar aA2ra
rAtat A2r A2ta aA2rt rAtar aA2a A3t aA3r rA2ta aAat A2tr tA3a rA3t
aAar Atra tA2at rA2tr aAta Atrt tA2ar rAtra aA2t At2r tAata rAtrt aAtr
Ar2a tAa2t rAt2r aAra Arat tAatr rAr2a aArt Arar tAara rArat aA2r Arta
tAart rArar tA2a Ar2t tAa2r rArta tAat Artr tAt2a rAr2t tAar A2ra tAtat
rArtr tAta A2rt tAtar rA2ra tA2t A3r tA2ta rA2rt tAtr aA3a tA3t rA3r

B. Secondary loop and whorl group:

R W. - to - R W

When no small letters are present, there are 9 possible combinations which can appear in the index fingers. They are as follows:

R U W - - - R R R

R U W - - - U U U

CHAPTER VI

R U W - - - W W W

At this point it is well to note that it may be preferable in some instances where small files are concerned to use only a portion of the classification formula in the filing sequence. In such cases, only those parts of the filing sequence which are necessary should be used along with the final and key.

III. Subsecondary:

III OOO. --- to --- III OOO

The sequence of the subsecondary is as follows:

III IIM IIO IMI IMM IMO IOI --- --- --- --- --- --- III III III III III III III

IOM IOO MII MIM MIO MMI MMM --- --- --- --- --- --- --- III III III III III III III

MMO MOI MOM MOO OII OIM OIO --- --- --- --- --- --- --- III III III III III III III

OMI OMM OMO OOI OOM OOO OOO, --- --- --- --- --- --- etc., to --- III III III III III III III OOO

each numerator in turn becoming the denominator for the complete sequence of numerators as listed above.

IV. Major:

The following sequence is used when loops appear in both thumbs:

S M L S M L S M L - - - - - - - - - S S S M M M L L L

When whorls appear in both thumbs the sequence is:

I M O I M O I M O - - - - - - - - - I I I M M M O O O

CHAPTER VI

When a whorl appears in the right thumb and a loop in the left, the sequence is:

I M O I M O I M O - - - - - - - - - S S S M M M L L L

When a loop appears in the right thumb and a whorl in the left, the sequence is:

S M L S M L S M L - - - - - - - - - I I I M M M O O O

V. Second Subsecondary:

SSS LLL. --- to --- SSS LLL

The sequence for filing the second subsecondary is as follows:

SSS SSM SSL SMS SMM SML --- --- --- --- --- --- SSS SSS SSS SSS SSS SSS

SLS SLM SLL MSS MSM MSL --- --- --- --- --- --- SSS SSS SSS SSS SSS SSS

MMS MMM MML MLS MLM MLL --- --- --- --- --- --- SSS SSS SSS SSS SSS SSS

LSS LSM LSL LMS LMM LML --- --- --- --- --- --- SSS SSS SSS SSS SSS SSS

LLS LLM LLL, LLL, --- --- --- etc., to --- SSS SSS SSS LLL

each group of the numerator becoming in turn the denominator for the complete sequence of numerators as listed above.

VI. W C D X Extensions:

W xX3x. - to ---- W xX3x

The sequence is as follows: Prints with c, d, or x in any finger other than the index fingers constitute the small-letter group. A sample of the

CHAPTER VI

sequence follows:

W cWc xWd Wdx cW cWd xWx Wxc dW cWx W2c Wxd xW dWc Wcd W2x Wc dWd Wcx cW2c Wd dWx Wdc cWcd Wx xWc W2d cWcx

As may be readily seen, the sequence proceeds in the same fashion as the a, t, r, small-letter sequence.

VII. Special Loop Extension used by the Federal Bureau of Investigation:

111 777. --- to --- 111 777

The following is a partial sequence for filing this extension:

111 112 113 114 115 116 117 --- --- --- --- --- --- --- 111 111 111 111 111 111 111

121 122 123 124 125 126 127 --- --- --- --- --- --- --- 111 111 111 111 111 111 111

131 132 133 134 135 136 137 --- --- --- --- --- --- --- 111 111 111 111 111 111 111

141 142 143 144 145 146 147 --- --- --- --- --- --- --- 111 111 111 111 111 111 111

151 152 153 154 155 156 157 --- --- --- --- --- --- --- 111 111 111 111 111 111 111

161 162 163 164 165 166 167 --- --- --- --- --- --- --- 111 111 111 111 111 111 111

171 172 173 174 175 176 177 777. --- --- --- --- --- --- --- etc., to --- 111 111 111 111 111 111 111 777

No matter how many of these divisions may be used, the order should remain the same; and no matter how many of these divisions are used, each individual group should be sequenced by:

VIII. Final:

Filed in numerical sequence from 1 out. For example, assume that there are 15 prints in a group having a final of 14. All of these should be filed together and followed by those prints in the same group having a final of 15, etc.

IX. Key:

All prints appearing in a designated final group are arranged by key in numerical sequence from 1 out. For example, assume that there are 5 prints in a group having a key of 14. All of these should be filed together and followed by those prints in the same group having a key of 15, etc.

CHAPTER VII

Searching and Referencing

Searching

When searching a print through the fingerprint files in order to establish an identification, it should be remembered that the fingerprint cards are filed in such a way that all those prints having the same classification are together. Thus, the print being searched is compared only with the groups having a comparable classification, rather than with the whole file.

After locating the proper group classification, the searcher should fix in his mind the one or two most outstanding characteristics of the patterns of the current print and look for them among the prints in file. If a print is found which has a characteristic resembling one upon the current print, the two prints should be examined closely to determine if identical. To avoid making an erroneous identification, the searcher should be exceedingly careful to ascertain that the prints being compared are identical in all respects before identifying one against the other.

To establish identity, it is necessary to locate several points of identity among the characteristics of the prints. The number of identical characteristics is left to the discretion of the individual but he should be absolutely certain that the prints are identical before treating them as such. Characteristics need not appear within the pattern area, since any ridge formation is acceptable. Quite often excellent ridge detail appears in the second joint of the finger. The characteristics used to establish an identification are shown in figure 102.

The final and the key may be considered control figures for searching prints. They limit the number of prints it is necessary to search in a group to those prints having finals and keys closely related to the final and key of the print being searched.

Due to the possibility of visual misinterpretation, distortion by pressure, or poor condition of the ridge detail of the prints in file, it is advisable to allow a margin for such discrepancies. Except in cases where the ridge count of the final and/or key is questionable on the print being searched, the following procedure is used:

Of the prints within any group classification, only those prints are examined which have a final within 2 ridge counts on each side of the final of the print being searched. For example, if the print to be searched has a final of 17, all prints bearing a final 15 through 19 will be compared with it.

Within the final of any group classification, only those prints are examined which have a key within 2 ridge counts on each side of the key of the print being searched. For example, if the print to be searched has a key of 20, all prints bearing a key of 18 through 22 will be compared with it.

In figure 352, it will be noted that there are 17 ridge counts appearing in the right little finger and this number is used as the final. It will also be noted that there is a loop of 24 ridge counts in the right thumb and this number is used as the key inasmuch as it is the first loop. In this example, the print is searched in the group classification which has finals ranging from 15 through 19. Within this group of finals the prints which have keys ranging from 22 through 26 are examined.

Referencing

Too much stress cannot be placed upon the necessity of referencing questionable patterns, whether it be in the interpretation of the type of pattern, the ridge count, or the tracing.

The factors which make it necessary are: variation in individual judgment and eyesight, the amount of ink used, the amount of pressure used in taking the prints, the difference in width of the rolled impressions, skin diseases, worn ridges due to age or occupations, temporary and permanent scars, bandaged fingers, crippled hands, and amputation.

For the highest degree of accuracy, all rolled impressions should be checked by the plain impressions, which generally are not distorted by

CHAPTER VII   91

pressure. This also helps prevent error caused by the reversal or mixing of the rolled impressions out of their proper order. For the same reason, as much of the counting and tracing should be done in the plain impressions as it is possible to do.

If there is any doubt as to which of two or more classifications should be assigned to a given pattern, it is given the preferred classification and reference searches are conducted in all other possible classifications. For example, if on a print with the preferred classification

1 A ---- 1 Aa

it is questionable whether the left middle finger should be a plain arch, a tented arch, or a radial loop, the print is searched in the

1 A ---- 1 Aa

group, and reference searches are conducted in the

1 A ---- 1 At

and

1 A ---- 1 Ar

groups. For further illustration, a print is given a preferred primary classification of

1, - 1

although the ridge detail on the right thumb is so formed as to resemble a whorl. The search is completed first in the preferred

1 - 1

primary classification and a reference search is then conducted in the

1 -- 17

primary.

All ridge counts that are "line counts," i.e., when one more or one less count would change the designation of the loop from I to O or from S to M, etc., must be searched in both groups. For example, in a print classified

16 M 1 U III 10, --------------- M 1 U III

if the ridge count of the right middle finger is 10 and the count in the right thumb is 16 (as indicated by the key), the print would be searched first as classified, then reference searches would be conducted in the following groups:

M 1 U IOI, L 1 U III, L 1 U IOI --------- --------- and --------- M 1 U III M 1 U III M 1 U III

When there is doubt concerning the tracing of a whorl, it should be treated in the same fashion. For example, if in the classification

O 5 U ------ I 17 U

doubt existed as to whether the tracing of the right thumb might not be a meeting tracing, the print would be searched as classified, and a reference search would be conducted in

M 5 U. ------ I 17 U

If there is no doubt concerning the ridge count used for the final, it is enough to search out of the group only those prints containing a final within 2 ridge counts on each side of the final on the print being searched. When, however, there is doubt concerning the ridge count of the final, the print should be searched 2 ridge counts on each side of the two extremes of possibility. For example, if it were possible for a final to be 6, 7, 8, or 9 ridge counts, the print should be searched through that part of the group bearing finals of from 4 through 11.

The above explanation pertaining to the final also applies to the key.

CHAPTER VII

All prints bearing amputations should be referenced to the necessary files containing prints other than amputations for reference searches.

In instances where only one finger is amputated, reference searches are conducted in all possible classifications, including all possible ridge counts or tracings. For example, a print containing the classification:

AMP

4 S 1 U III 6 -------------- S 1 U III

with the right index finger amputated, the left index finger being an ulnar loop, would be searched first in the amputation group for the classification, then reference searches would be conducted in the following groups in the nonamputation files:

S 1 U III S 1 T II S 17 W III --------- --------- ---------- S 1 U III S 1 U III S 1 U III

S 1 U OII S 1 R III S 17 W MII --------- --------- ---------- S 1 U III S 1 U III S 1 U III

S 1 A II S 1 R OII S 17 W OII --------- --------- ---------- S 1 U III S 1 U III S 1 U III

All prints bearing unprinted or badly crippled fingers are filed in the nonamputation files, and reference searches are conducted in the amputation group.

For the purpose of determining if it is feasible to conduct reference searches in all possible classifications, the method of referencing amputations is applied to completely scarred patterns (Chapter titled "Scarred Patterns--Amputations--Missing at Birth"). For example, a print bearing the preferred classification:

13 O 17 W OOO 14 ---------------- L 17 U OOI

with the left middle finger completely scarred, the right middle finger being an ulnar loop with a ridge count of 13, would be searched first in the group for that classification, then reference searches would be conducted in the following groups:

O 17 W OOO O 17 W O 19 W OOO ---------- ------- ---------- L 17 U OII L 17 Ur L 17 U OOI

O 17 W O 19 W OOO ------- ---------- L 17 Ua L 17 U OII

O 17 W O 19 W OOO ------- ---------- L 17 Ut L 17 U OMI

The referencing of partial scars is a problem in which many factors are present. A full explanation of the scars, their preferred classifications and their references is made in the chapter, "Classification of Scarred Patterns--Amputations--Missing at Birth."

When the age extension is utilized and a "Reference" group and a "Presumptive Dead" file are maintained, it is suggested that a general allowance of 5 years be considered to allow for a discrepancy in prints bearing the ages of 50 years or older.

In the files of the Federal Bureau of Investigation the various age groups are as follows:

1-54 "Regular" file. 55-74 "Reference" file. 75 and over "Presumptive Dead" file.

Reference searches for the preceding groups are conducted in the following manner:

50-54 Referenced to "Reference" file. 70-74 Referenced to "Presumptive Dead" file and "Regular" file. 75-79 Referenced to "Reference" file and "Regular" file. 80 and older Referenced to "Regular" file only.

If no age is given, it should be searched first in the regular file and reference searches should be conducted in the "Reference" group and the "Presumptive Dead" file.

CHAPTER VII

When separate male and female files are maintained, there may be doubt as to the sex of a subject due to a discrepancy between the sex indicated and the name and the description and picture. In such case try to determine the sex from the description and the size of the prints, then reference the print to the other file. A Photostat copy can be made and placed in the other file until the true sex can be determined.

CHAPTER VIII

How To Take Inked Fingerprints

The equipment required for taking fingerprints consists of an inking plate, a cardholder, printer's ink (heavy black paste), and a roller. This equipment is simple and inexpensive.

In order to obtain clear, distinct fingerprints, it is necessary to spread the printer's ink in a thin even coating on a small inking plate. A roller similar to that used by printers in making galley proofs is best adapted for use as a spreader. Its size is a matter determined by individual needs and preferences; however, a roller approximately 6 inches long and 2 inches in diameter has been found to be very satisfactory. These rollers may be obtained from a fingerprint supply company or a printing supply house.

[Illustration: 360. Fingerprint stand.]

An inking plate may be made from a hard, rigid, scratch-resistant metal plate 6 inches wide by 14 inches long or by inlaying a block of wood with a piece of glass one-fourth of an inch thick, 6 inches wide, and 14 inches long. The glass plate by itself would be suitable, but it should be fixed to a base in order to prevent breakage. The inking surface should be elevated to a sufficient height to allow the subject's forearm to assume a horizontal position when the fingers are being inked. For example, the inking plate may be placed on the edge of a counter or a table of counter height. In such a position, the operator has greater assurance of avoiding accidental strain or pressure on the fingers and should be able to procure more uniform impressions. The inking plate should also be placed so that the subject's fingers which are not being printed can be made to "swing" off the table to prevent their interfering with the inking process. A fingerprint stand such as that shown in figure 360 may be purchased from fingerprint supply companies. The stand is made of hardwood and measures approximately 2 feet in length, 1 foot in height and width. This stand contains a cardholder and a chrome strip which is used as the inking plate. Two compartments used to store blank fingerprint cards and supplies complete the stand. This equipment should be supplemented by a cleansing fluid and

CHAPTER VIII

necessary cloths so that the subject's fingers may be cleaned before rolling and the inking plate cleaned after using. Denatured alcohol and commercially available cleaning fluids are suitable for this purpose.

[Illustration: 361. Fingerprints properly taken.]

PERSONAL	ROE	RICHARD	RANDOLPH	SEX	IDENTIFICATION	
	LAST NAME	FIRST NAME	MIDDLE NAME	MALE		
FINGERPRINTS SUBMITTED BY				RACE		W
				HT. (Inches)	WT.	
SIGNATURE OF PERSON FINGERPRINTED				71	170	
						1655 Grant Avenue
DATE OF BIRTH					6/6/42	
FINGERPRINTED BY						Chicago, Illinois
				HAIR	EYES	RESIDENCE OF PERSON FINGERPRINTED
				BR	BR	
DATE FINGERPRINTED					8/12/62	LEAVE THIS SPACE BLANK
PERSON TO BE NOTIFIED IN CASE OF EMERGENCY					Thomas L. Roe	CLASS NAME
PLACE OF BIRTH					Omaha, Neb.	
ADDRESS				1655 Grant Avenue		
CITIZENSHIP				Chicago, Illinois	American	
REF.						
SCARS AND MARKS				Appendectomy		See Reverse Side for Further Instructions

1. RIGHT THUMB	2. RIGHT INDEX	3. RIGHT MIDDLE	4. RIGHT RING	5. RIGHT LITTLE
[Illustration]	[Illustration]	[Illustration]	[Illustration]	[Illustration]

1. LEFT THUMB	2. LEFT INDEX	3. LEFT MIDDLE	4. LEFT RING	5. LEFT LITTLE
[Illustration]	[Illustration]	[Illustration]	[Illustration]	[Illustration]

LEFT FOUR FINGERS TAKEN SIMULTANEOUSLY	LEFT THUMB	RIGHT THUMB	RIGHT FOUR FINGERS TAKEN SIMULTANEOUSLY
[Illustration]	[Illustration]	[Illustration]	[Illustration]

CHAPTER VIII

The fingerprints should be taken on 8- by 8-inch cardstock, as this size has generally been adopted by law enforcement because of facility in filing and desirability of uniformity. Figure 361 shows fingerprints properly taken on one of the standard personnel identification cards from the Federal Bureau of Investigation. From this illustration, it is evident there are two types of impressions involved in the process of taking fingerprints. The upper 10 prints are taken individually--thumb, index, middle, ring, and little fingers of each hand in the order named. These are called "rolled" impressions, the fingers being rolled from side to side in order to obtain all available ridge detail. The smaller impressions at the bottom of the card are taken by simultaneously printing all of the fingers of each hand and then the thumb without rolling. These are called "plain" or "fixed" impressions and are used as a check upon the sequence and accuracy of the rolled impressions. Rolled impressions must be taken carefully in order to insure that an accurate fingerprint classification can be obtained by examination of the various patterns. It is also necessary that each focal point (cores and all deltas) be clearly printed in order that accurate ridge counts and tracings may be obtained.

In preparing to take a set of fingerprints, a small daub of ink should be placed on the inking glass or slab and thoroughly rolled until a very thin, even film covers the entire surface. The subject should stand in front of and at forearm's length from the inking plate. In taking the rolled impressions, the side of the bulb of the finger is placed upon the inking plate and the finger is rolled to the other side until it faces the opposite direction. Care should be exercised so the bulb of each finger is inked evenly from the tip to below the first joint. By pressing the finger lightly on the card and rolling in exactly the same manner, a clear rolled impression of the finger surface may be obtained. It is better to ink and print each finger separately beginning with the right thumb and then, in order, the index, middle, ring, and little fingers. (Stamp pad ink, printing ink, ordinary writing ink, or other colored inks are not suitable for use in fingerprint work as they are too light or thin and do not dry quickly.)

If consideration is given the anatomical or bony structure of the forearm when taking rolled impressions, more uniform impressions will be obtained. The two principal bones of the forearm are known as the

CHAPTER VIII   99

radius and the ulna, the former being on the thumb side and the latter on the little finger side of the arm. As suggested by its name, the radius bone revolves freely about the ulna as a spoke of a wheel about the hub. In order to take advantage of the natural movement in making finger impressions, the hand should be rotated from the awkward to the easy position. This requires that the thumbs be rolled toward and the fingers away from the center of the subject's body. This process relieves strain and leaves the fingers relaxed upon the completion of rolling so that they may be lifted easily from the card without danger of slipping which smudges and blurs the prints. Figures 362 and 363 show the proper method of holding a finger for inking and printing a rolled impression.

The degree of pressure to be exerted in inking and taking rolled impressions is important, and this may best be determined through experience and observation. It is quite important, however, that the subject be cautioned to relax and refrain from trying to help the operator by exerting pressure as this prevents the operator from gaging the amount needed. A method which is helpful in effecting the relaxation of a subject's hand is that of instructing him to look at some distant object and not to look at his hands. The person taking the fingerprints should stand to the left of the subject when printing the right hand, and to the right of the subject when printing the left hand. In any case, the positions of both subject and operator should be natural and relaxed if the best fingerprints are to be obtained.

To obtain "plain" impressions, all the fingers of the right hand should be pressed lightly upon the inking plate, then pressed simultaneously upon the lower right hand corner of the card in the space provided. The left hand should be similarly printed, and the thumbs of both hands should be inked and printed, without rolling, in the space provided. Figures 364 and 365 show the correct method of taking plain impressions of the fingers and thumbs.

[Illustration: 362. Proper method of holding finger.]

[Illustration: 363. Proper method of printing rolled impressions.]

[Illustration: 364. Proper method of taking plain impressions of fingers.]

CHAPTER VIII

[Illustration: 365. Proper method of taking plain impressions of thumbs.]

CHAPTER IX

Problems in the Taking of Inked Fingerprints

From time to time various problems arise concerning the taking of inked impressions. It is believed that these problems can be divided into four phases:

- Mechanical operation

- Temporary disabilities

- Permanent disabilities

- General

Mechanical operation

In order to take good fingerprints, the necessary equipment should be maintained in a neat and orderly manner at all times.

Poor impressions are usually caused by one of the following faults:

1. The use of poor, thin, or colored ink, resulting in impressions which are too light and faint, or in which the ink has run, obliterating the ridges. The best results will be obtained by using heavy black printer's ink, a paste which should not be thinned before using. This ink will dry quickly and will not blur or smear with handling.

2. Failure to clean thoroughly the inking apparatus and the fingers of foreign substances and perspiration, causing the appearance of false markings and the disappearance of characteristics. Windshield cleaner, gasoline, benzine, and alcohol are good cleansing agents, but any fluid may be used. In warm weather each finger should be wiped dry of perspiration before printing.

3. Failure to roll the fingers fully from one side to the other and to ink the whole area from tip to below the first fissure. The result of this is that the focal points of the impressions (the deltas or cores) do not

CHAPTER IX
102

appear. The whole finger surface from joint to tip and from side to side should appear.

4. The use of too much ink, obliterating or obscuring the ridges. If printer's ink is used, just a touch of the tube end to the inking plate will suffice for several sets of prints. It should be spread to a thin, even film by rolling.

5. Insufficient ink, resulting in ridges too light and faint to be counted or traced.

6. Allowing the fingers to slip or twist, resulting in smears, blurs, and false-appearing patterns. The fingers should be held lightly without too much pressure. The subject should be warned not to try to help but to remain passive.

The illustrations numbered 366 through 377 show the results of these faults and show also the same fingers taken in the proper manner.

Illegible inked prints

A brief review of the problems of classifying and filing a fingerprint card in the FBI will help to clarify the FBI's policy concerning the processing of "bad" inked fingerprints.

The criminal fingerprint file contains the fingerprints of millions of individuals. The complete classification formula is used. To obtain it, each inked finger must show all the essential characteristics. Because of the immense volume of prints it has become necessary to extend the normal classification formula.

To illustrate this point:

dWdwc xCdwc O 32 W OOO 18 I 32 W III

In order to subdivide the 32 over 32 primary still further, the ridge count of the whorl of the right little finger is used to obtain a final classification. The extension above the normal classification formula indicates that each whorl is classified as to the type; namely, plain

CHAPTER IX 103

whorl (W), double loop (D), central pocket loop (C), and accidental (X). Accordingly, it is not enough for the FBI Identification Division to ascertain the general whorl pattern type, but the deltas and core must show in order to obtain the ridge tracing, the type of whorl, and also, in some instances, the ridge count. The complete WCDX extension is outlined in Chapter VI.

Figures 366 to 377 are some examples of improperly and properly taken inked fingerprints.

An examination of figure 372 shows that it is a whorl. In order to classify the ridge tracing accurately, however, so that the fingerprint card can be placed in the correct classification, the left delta must show. The approximate ridge tracing for the whorl in figure 372 would be MEETING. An examination of the properly taken fingerprint in figure 373 indicates that the correct ridge tracing is INNER. It follows that the pattern in figure 372 would not have been placed in the proper place in file.

[Illustration: 366. Improper.]

[Illustration: 367. Proper.]

[Illustration: 368. Improper.]

[Illustration: 369. Proper.]

[Illustration: 370. Improper.]

[Illustration: 371. Proper.]

[Illustration: 372. Improper.]

[Illustration: 373. Proper.]

The correct whorl tracing is needed to obtain the complete subsecondary and the major classifications.

CHAPTER IX

It may be noted that both deltas are present in figure 374. This would enable the technical expert to ascertain the correct ridge tracing, OUTER. In the core of the whorl, however, there is a heavy amount of ink which makes it impossible to determine the type of whorl with any degree of accuracy. If one were to hazard a guess, it would appear to be a plain whorl. Actually, the correct type of whorl, a double loop, is clearly visible in figure 375.

It can be ascertained that the pattern in figure 376 is a loop, but an accurate ridge count cannot be obtained because the left delta does not appear. The approximate ridge count of this loop is 14 to 16. This approximation is sufficient for a fingerprint expert to place this loop in the "O" group of any finger of the subsecondary. The correct ridge count of this loop is 19, and it appears in illustration 377. The approximate ridge count is not sufficient to place this print properly in the large files of the FBI because in certain general complete classification formulas the accurate ridge count is needed to obtain an extension. These extensions use a smaller grouping of ridge counts to form a valuation table, and in this way, differ from the larger grouping of ridge counts which form the basis of the subsecondary classification. These extensions are called the second subsecondary and the special loop extension and are outlined in chapter VI.

[Illustration: 374. Improper.]

[Illustration: 375. Proper.]

[Illustration: 376. Improper.]

[Illustration: 377. Proper.]

There are two additional points which illustrate the FBI's need for the delta, ridges, and core to show clearly in loops. The first point is set forth: the ridge count of the loop may be needed to obtain the key classification. The key classification is an actual ridge count, and no valuation table is used to obtain a subdivision. The key classification is used as an integral part of the fingerprint filing system. The second point is as follows: the ridge count may be needed to obtain the final classification. The final classification is an actual ridge count, and no valuation table is used to obtain a subdivision. The final classification is used as an integral part of the fingerprint filing system.

The following are just a few examples to illustrate the completeness of the classification formula used in the FBI fingerprint file:

12 M 9 R OIO 11

S 1 R IOI

Key Major Primary Secondary Subsecondary Final

6 17 aW IIO 9

1 U OII Key Primary Small letter Subsecondary Final Secondary

8 S 1 Ua II 6 S 1 U III

Key Major Primary Small letter (Subsecondary Final Secondary Extension)

SML (Second SML Subsecondary) 5 0 5 U IOO 14

I 17 U IOO

Key Major Primary Secondary Subsecondary Final

245 (Special Loop 332 Extension) 14 M 1 U IOO 16

S 1 U OII

Key Major Primary Secondary Subsecondary Final

15 I 29 W IOO 19

I 28 W OOI

Key Major Primary Secondary Subsecondary Final

CHAPTER IX 106

These several examples should help to illustrate the FBI's extended classification formulas for classifying and filing fingerprints. The larger collection of fingerprints must of necessity call for a more detailed analysis of all fingerprint characteristic details. The closer examination to obtain further fingerprint subdivisions is dependent on ten legible inked impressions.

The identification officer will understand the problems of accurately classifying and filing fingerprint cards. He knows there is little value in placing a fingerprint card in the FBI's files with only an approximate or an inaccurate classification.

Every fingerprint card filed in the FBI's file is of value to the particular law enforcement agency which forwarded it, as well as to all other law enforcement agencies which rely on its being correctly classified and filed.

Temporary disabilities

There are temporary disabilities affecting an individual's hand which are sometimes beyond the control of the identification officer. These can be fresh cuts, or wounds, bandaged fingers or finger, occupational (carpenters, bricklayers, etc.) blisters, and excessive perspiration. Children, whose ridges are small and fine, would also come under this heading. Extreme care should be exercised in fingerprinting the aforementioned.

An indication on the fingerprint card to the effect "fresh cut, bandaged" is not sufficient to file the fingerprint card. It is obvious that a fingerprint card bearing these notations cannot be properly classified and filed. The same situation would occur if there were a blister on an individual's finger. The blister temporarily disfigures the ridge detail. When an injury is temporary, the prints, if at all possible, should not be taken until after the injury has healed.

Occupational problems (bricklayers, carpenters, etc.) are definitely a challenge to the identification officer. In some instances, by means of softening agents (oils and creams), it is possible to obtain legible inked impressions. It is further suggested that in these cases a very small amount of ink should be used on the inking plate.

Excessive perspiration can be controlled to some extent by the identification officer. Excessive perspiration causes the inked impressions to be indistinct. It is suggested in these cases to wipe the finger with a cloth and then immediately ink the finger and roll it on the fingerprint card. This process should be followed with each finger. It is also suggested that possibly the fingers could be wiped with alcohol, benzine, or similar fluid which would act as a drying agent.

In all the above situations, if it is not possible to accurately classify and file the fingerprint card, the name appearing on the card will be searched in the alphabetical files and then returned to the law-enforcement agency.

Permanent disabilities

Another phase involves permanent disabilities which can in most cases be controlled by the identification officer. These can be lack of fingers (born without), amputations, crippled fingers (bent, broken), deformities (webbed, extra fingers), and old age.

With respect to lack of fingers, it should be noted that some individuals are born without certain fingers. The notation "missing" is not satisfactory because it does not sufficiently explain the correct situation. It is suggested that "missing at birth" or some similar notation be made in the individual fingerprint block on the card. A proper notation concerning this situation will prevent the fingerprint card from being returned. Figures 378 and 379 illustrate temporary and permanent disabilities.

[Illustration: 378. Temporary disability.]

[Illustration: 379. Permanent disability.]

Concerning amputations, it is suggested that a proper notation to this effect appear in the individual fingerprint block or blocks. It is suggested that if a portion of the first joint of a finger is amputated, the finger should be inked and printed. A notation concerning this fact should be made on the fingerprint card in the individual fingerprint block.

In those cases where all of the fingers are amputated, the inked footprints should be obtained.

The handling of crippled fingers and certain deformities can be discussed in a group because they generally present the same problems. It is not sufficient in all cases to indicate "broken," "bent," "crippled." If the fingers are bent or crippled so that they are touching the palm and cannot be moved, a notation to this effect should be on the fingerprint card in the proper individual fingerprint block. However, it is believed that these extreme cases are rare. It is suggested that the special inking devices used for taking the prints of deceased individuals be used in taking inked impressions of bent or crippled fingers.

[Illustration: 380. The spatula, roller, and curved holder used for taking the inked prints of bent or crippled fingers.]

This equipment, which will be discussed more fully in the section on printing deceased persons, consists of spatula, small roller, and a curved holder for the individual finger block. Figure 380 shows the spatula, roller, and curved holder. It should be further noted in figure 380 that there are a strip of the entire hand of the fingerprint card and also individual finger blocks cut from the fingerprint card. Each of these types can be used in connection with the curved holder.

Each crippled finger is taken as a separate unit and then the finger block pasted on a fingerprint card. In figure 381, note the use of the spatula for applying the ink to a bent or crippled finger; and in figure 382, observe the use of the curved holder for taking the "rolled" impression of a bent or crippled finger.

Old age has been placed under permanent disability only for discussion purposes. The problem is not encountered frequently in taking the fingerprints of individuals who are arrested. The situation of crippled fingers due to old age may be met, and it can be handled as previously suggested. In most cases the problems arise because of the very faint ridges of the individual. It is believed that in the majority of cases, legibly inked prints can be taken by using a very small amount of ink on the inking plate and by using little pressure in the rolling of the fingers.

[Illustration: 381. The use of the spatula in the application of ink to the finger.]

[Illustration: 382. The use of the curved holder for taking the "rolled" impression.]

Deformities

If a subject has more than 10 fingers, as occasionally happens, the thumbs and the next 4 fingers to them should be printed, and any fingers left over should be printed on the other side of the card with a notation made to the effect that they are extra fingers. When a person with more than 10 fingers has an intentional amputation performed, it is invariably the extra finger on the little finger side which is amputated.

[Illustration: 383]

It also happens, not infrequently, that a subject will have two or more fingers webbed or grown together, as in figures 383 and 384, making it impossible to roll such fingers on the inside. Such fingers should be rolled, however, as completely as possible, and a notation made to the effect that they are joined.

Split thumbs, i.e., thumbs having two nail joints, as in figure 385, are classified as if the joint toward the outside of the hand were not present. In other words the inner joint is used, and no consideration whatever is given to the outer joint.

[Illustration: 384]

[Illustration: 385]

General

CHAPTER IX

These problems have dealt with the mechanical or operational processes. However, there are other problems dealing with the completing of the descriptive information. The fingerprint card may be returned because of the lack of information in the spaces provided, such as name, sex, race, height, weight, etc. Any discrepancies in this information may necessitate the return of the fingerprint card.

The success and value of the FBI's fingerprint files to all law enforcement agencies are dependent, in a large measure, on the legibly inked fingerprints taken by law enforcement agencies.

Figure 386 shows an enlarged portion of the bulb of a finger revealing the microscopic structure of the friction skin. The epidermis consists of two main layers, namely, the stratum corneum, which covers the surface, and the stratum mucosum, which is just beneath the covering surface. The stratum mucosum is folded under the surface so as to form ridges which will run lengthwise and correspond to the surface ridges. However, these are twice as numerous since the deeper ridges which correspond to the middle of the surface ridges alternate with smaller ones which correspond to the furrows. The sweat pores run in single rows along the ridges and communicate through the sweat ducts with the coil sweat glands which are below the entire epidermis. The friction ridges result from the fusion in rows of separate epidermic elements, such as the dot shown on the left. Generally speaking, when an individual bruises or slightly cuts the outer layer or stratum corneum of the bulb of the finger, the ridges will not be permanently defaced. However, if a more serious injury is inflicted on the bulb of the finger, thereby damaging the stratum mucosum, the friction skin will heal, but not in its original formation. The serious injury will result in a permanent scar appearing on the bulb of the finger.

[Illustration: 386]

CHAPTER X

Problems and Practices in Fingerprinting the Dead

Each year new graves are opened in potter's fields all over the United States. Into many of them are placed the unknown dead--those who have lived anonymously or who, through accident or otherwise, lose their lives under such circumstances that identification seems impossible. In a majority of such cases, after the burial of the body, no single item or clue remains to effect subsequent identification. As a result, active investigation usually ceases and the cases are forgotten, unless, of course, it is definitely established that a murder has been committed.

Reliance is too often placed on visual inspection in establishing the identity of the deceased. This includes having the remains viewed by individuals seeking to locate a lost friend or relative. The body is often decomposed. If death was caused by burning, the victim may be unrecognizable. As a result of many fatal accidents the deceased is often mutilated, particularly about the face, so that visual identification is impossible. Yet, in many cases, the only attempt at identification is by having persons view the remains and the personal effects.

The recorded instances of erroneous visual identifications are numerous. In one case a body, burned beyond recognition, was identified by relatives as that of a 21-year-old man; yet fingerprints later proved that the corpse was that of a 55-year-old man.

Fingerprints have frequently been instrumental in establishing the correct identity of persons killed in airplane crashes and incorrectly "identified" by close relatives.

In one instance a woman found dead in a hotel room was "positively" identified by several close friends. The body was shipped to the father of the alleged deceased in another state where again it was "identified" by close friends. Burial followed. Approximately one month later the persons who had first identified the body as that of their friend were sitting in a tavern when the "dead" woman walked into the room. Authorities were immediately advised of the error; they in turn advised

CHAPTER X

the authorities in the neighboring state of the erroneous identification and steps were taken immediately to rectify the mistake. After permission had been granted by the State Health Board to exhume the body of the dead woman, fingerprints were taken and copies were forwarded to the FBI Identification Division. The finger impressions were searched through the fingerprint files and the true identity of the deceased was established.

During a 12-month period, the FBI Identification Division received the fingerprints of 1,708 unknown dead. Of these, 1,298, or almost 76 percent, were identified. The remaining 410 were not identified simply because fingerprints of these individuals were not in the FBI files. It should be noted that in these 1,708 cases, it was possible to secure legible fingerprints of the deceased in the usual manner by inking the fingers in those instances in which decomposition had not injured the ridge detail.

[Illustration: 387. Field equipment for disaster identification.]

In addition to the fingerprints of 1,708 unknown dead, the Identification Division received the fingers and/or the hands of 85 unknown dead individuals. In these cases, decomposition was so far advanced that it was not possible to secure inked fingerprints in the regular manner. Of these, 68 bodies, or 80 percent of the group, were identified. Of the 17 unidentified, the fingerprints of 14 were not in the FBI files. In three cases decomposition was so far advanced that all ridge detail had been destroyed.

In order to emphasize what can be accomplished, it is pointed out that in those cases in which hands and fingers were submitted, the time which elapsed from death until the specimens were received ranged from a week to 3 years. Incredible as it may seem, it has been possible to secure identifiable impressions 3 years after death.

These statistics of achievement in the field of identifying unknown dead re-emphasize the fact that in all cases involving the identification of a deceased person, fingerprints should be used as the medium for establishing a conclusive and positive identification.

Generally speaking, in the course of their work fingerprint operators find it necessary to take the impressions of three classes of deceased persons.

They are:

- Those who have died recently, in which cases the task is relatively simple.

- Those dead for a longer period, in which cases difficulty is experienced due to pronounced stiffening of the fingers, the early stages of decomposition, or both.

- Those cases in which extreme difficulty is encountered because of maceration, desiccation, or advanced decay of the skin.

These problems will be considered separately.

1. Fingerprinting the Newly Dead.

When the fingers are flexible it is often possible to secure inked fingerprint impressions of a deceased person through the regular inking process on a standard fingerprint card. Experience has proved that this task can be made easier if the deceased is laid face down and palms down on a table (fig. 388).

In all cases where inked impressions are to be made, care should be exercised to see that the fingers are clean and dry before inking. If necessary, wash the digits with soap and water and dry thoroughly.

In the event difficulty is encountered in trying to procure fingerprints by the regular method, it may prove more convenient to cut the 10 "squares" numbered for the rolled impressions from a fingerprint card. After the finger is inked, the square is rolled around the finger without letting it slip. Extreme caution should be exercised to see that each square bears the correct fingerprint impression. After all the inked impressions are properly taken, the ten squares bearing the impressions are pasted or stapled to a standard fingerprint card in their proper positions, i.e., right thumb, right index, right middle, etc.

Whenever possible the "plain" or "simultaneous" impressions should also be taken.

In some cases it will be found necessary to obtain or improvise a tool similar to a broad-bladed putty knife or spatula to be used as an inking instrument. The ink is rolled evenly and thinly on the knife or spatula and applied to the finger by passing the inked knife or spatula around it. The tool, of course, replaces the usual glass inking slab or plate, the use of which is extremely difficult or awkward when printing a deceased person.

2. Fingerprinting the Dead, Where Stiffening of the Fingers and/or Early Decomposition Are Present.

This second group consists of cases in which the hands of the deceased are clenched, or the finger tips are wrinkled, or decomposition has begun, and/or where there are combinations of these three conditions. Cases of this sort may necessitate cutting off the skin. Legal authority is necessary before cutting a corpse. Such authority may be granted by state law or by an official having authority to grant such a right.

[Illustration: 388]

In cases where rigor mortis (stiffening of the muscles) has set in and the fingers are tightly clenched, the fingers may be forcibly straightened by "breaking the rigor." This is done by holding the hand of the deceased person firmly with one hand, grasping the finger to be straightened with the four fingers of the other hand and placing the thumb, which is used as a lever, on the knuckle of the finger and forcing it straight (fig. 389). The inking tool and "squares," as previously explained, are then used to secure the fingerprint.

In the event the rigor cannot be completely overcome, it will be most helpful to improvise or secure a spoon-shaped tool for holding the cut squares or cut strips while printing the fingers, similar to the tool mentioned briefly in the discussion of crippled fingers. This tool, somewhat resembling a gouge without the sharp edge, should have a handle, a concave end, and a frame or clamp to hold the cardboard squares or strips. In Figure 390, one type of tool is illustrated. This tool eliminates the necessity of rolling the deceased's finger, since the "square" assumes the concave shape of the tool, and the gentle pressure applied to the inked finger when it is brought in contact with the square results in a "rolled" impression without actually rolling the finger.

[Illustration: 389]

Another problem encountered in this second group includes cases in which the tips of the fingers are fairly pliable and intact, yet due to the presence of wrinkles in the skin, complete impressions cannot be obtained. This condition can be corrected by the injection of a tissue builder, procurable from a dealer in undertaker's supplies. If this is not available, glycerin or water may be used.

The method is simple. Injection of the tissue builder, glycerin, or water, is accomplished by the use of a hypodermic syringe. The hypodermic needle is injected at the joint of the finger up into the tip of the finger, care being used to keep the needle below the skin surface (fig. 391). The solution is injected until the finger "bulbs" are rounded out, after which they are inked and printed.

Occasionally, in stubborn cases, entry of the needle at the joint and injection of the fluid will not completely fill the finger bulb. It may be necessary, therefore, to inject the fluid at other points of the finger such as the extreme tip or sides, until suitable results are achieved (fig. 392). The tissue builder has a distinct advantage over glycerin or water, inasmuch as the builder hardens after a short time and is not lost, whereas glycerin and water sometimes seep out when pressure is applied in printing. To offset seepage at the point where the hypodermic needle is injected, whenever possible, tie a piece of string tightly around the finger just above the point of entry of the needle.

When the tissue builder is purchased, a solvent for cleaning the hypodermic syringe and needle should be acquired, inasmuch as the builder will harden in the syringe and needle.

[Illustration: 390]

CHAPTER X

[Illustration: 391]

[Illustration: 392]

Those cases in which decomposition in its early stage is present belong in this group also. Frequently, the outer layer of skin has begun to peel from the fingers. A careful examination should be made to determine if the peeling skin is intact or if a part of it has been lost. If the skin is in one piece, an effort should be made to secure prints just as though it were attached normally to the finger. Or, if it is deemed advisable, the skin may be peeled off in one piece, placed over the finger of the operator, and inked and printed as though it were his own finger.

Occasionally the first layer of skin is missing. There remains the dermis or second layer of skin which is also of value for identification purposes. This second layer would be dealt with as though it were the outside skin, using the techniques described above. The ridge detail of the second layer of skin is less pronounced than that of the outer skin, however, and more attention and care are needed in order to obtain suitable impressions.

So far this discussion has dealt with the taking of impressions of fingers when the flesh is fairly firm and the ridge detail intact. A different problem arises when the fingers are in various stages of decay. The techniques of treating the fingers in such cases vary greatly, depending upon the condition of the fingers with respect to decomposition, desiccation, or maceration.

3. Fingerprinting the Dead in Difficult Cases.

In cases involving badly decomposed bodies the first thing to do is to examine the fingers to see if all are present. If they are not, an effort should be made to determine whether the missing finger or fingers or even a hand was amputated during the person's lifetime, or whether the loss was due to other causes such as destruction by animal or marine life. Deductions from this examination should be noted on the fingerprint record. This point is made in view of the fact that in the fingerprint files of the FBI and some police departments, the fingerprint cards reflecting amputations are filed separately. Noting amputations may lessen to a great extent a search through the fingerprint files.

In making the initial examination, attention should be given to the removal of dirt, silt, grease and other foreign matter from the fingers. Soap and water are good cleansing agents. So is xylene, a chemical which will readily clean grease and fatty matter from the fingers. Good results can be achieved by utilizing a child's soft-bristled toothbrush in cases where the skin is fairly firm. The brushing should be done lightly and the strokes should follow the ridge design in order to clean not only the ridges but the depressions as well. In the event that the skin is not firm enough to use the toothbrush, a cotton swab may be used. The fingers should be wiped very lightly with either soap and water or xylene, always following the ridge contours.

At this point the fingers are again examined to determine the condition they are in, based upon the circumstances in which the body was found. Study and actual experience have shown that there are three general types of conditions to be considered: Decomposition or putrefaction, prevalent in bodies found in brush or buried in earth; desiccation or mummification (that is, dried out), noted in bodies which have been found in the open (ridge detail not in contact with the ground) in dry protected places, or bodies subjected to severe heat; and the group involving maceration (water soaking), which ordinarily results from being immersed in water.

The degree of decomposition, desiccation, or maceration varies from a comparatively early stage to an extremely advanced stage. Accordingly, each case must be considered individually. For example, what is done successfully in one case of desiccation may not show favorable results in another. Hence, the techniques outlined below point out generally what can be done, and has been done, with success.

[Illustration: 393]

When a body is found, the hands usually will be tightly clenched. The first problem will be to straighten the fingers. If rigor mortis has set in and an effort to straighten the fingers as previously explained fails, the difficulty can be overcome easily. Using a scalpel, make a deep cut at the second joint on the inner side of each of the four fingers. They can now be straightened with the application of force (fig. 393). The thumb, if it is cramped or bent, can generally be straightened by making a deep cut between the thumb and the index finger. These incisions are made for the obvious purpose of examining the fingers to determine if there is any ridge detail. Before this fact can be definitely ascertained it may be necessary to cleanse the pattern areas with soap and water or xylene, as previously explained.

[Illustration: 394]

Advanced decomposition

If the case is one involving decomposition, the operator is confronted with the problem of dealing with flesh which is rotted or putrefied. The flesh may be soft or flabby and very fragile. If this is so, an examination is made of the finger tips to see if the outer skin is present. If the outer skin is present and intact, it may be possible, using extreme care, to ink and print in the regular manner. Sometimes, the outer skin, although present, will be too soft and fragile to ink and roll in the regular way. In such cases, when the ridge detail is discernible, the skin, if it is easily removed from the finger, or the finger itself may be cut off at the second joint and placed in a 10- to 15-percent solution of formaldehyde for approximately an hour to harden it. Skin placed in a formaldehyde solution usually turns a grayish white and becomes firm. However, it will be brittle and may split if not handled carefully. The skin is placed in the solution only until it hardens sufficiently, after which it is removed and carefully wiped dry with a piece of cloth. Then the skin, placed over the operator's own thumb or index finger and held in place by his other hand, is inked and rolled as though the operator were printing his own finger. If a legible print is not obtainable in this manner, the operator should examine the underside of the skin.

[Illustration: 395]

In many instances, where the ridge detail on the outer surface has been destroyed or is not discernible, the ridge detail is clearly visible on the underside (fig. 394). If this is the case, the skin is inverted (turned inside out) very carefully to prevent splitting or breaking and then is inked and printed in the usual way. It must be borne in mind, however, that when the underside of the skin is printed the resulting impression will be in reverse color and position; that is, the ink is actually adhering to what would be furrows of the pattern when viewed from the proper or outer side. If it is deemed inadvisable to try to invert or turn the skin inside out for fear of damaging it, a photograph of the inner ridge detail is made and the negative is printed to give an "as is" position photograph for proper classification and comparison purposes. In order to secure good photographs of the ridge detail it may be advisable to trim the skin, flatten it out between two pieces of glass, and photograph it in that position (fig. 395).

When the entire finger is placed in the solution during the hardening process, the skin, in absorbing the formalin solution, may swell and come loose from the finger. Should this occur, the skin must be removed carefully and the procedure outlined above followed. If, however, the skin still adheres to the finger and is not too wrinkled, ink is applied and prints made. Should the skin be too wrinkled to secure a satisfactory impression, consideration is given to the injection of the tissue builder under the skin as previously mentioned, in order to distend the pattern area. If successful, the finger is inked and printed. This, of course, can be done only when the skin is intact.

Should part of the skin be destroyed to the extent that tissue builder cannot be injected effectively, while examination discloses that the pattern area is present but wrinkled, cut off the entire pattern area from the joint to the tip of the finger (fig. 396). Care must be exercised to insure getting the complete fingerprint pattern as well as cutting deep enough to avoid injury to the skin.

[Illustration: 396]

After excision, the flesh is carefully and meticulously removed from the inside by scraping, cutting, and trimming until only the skin remains, or until the specimen is so thin it can be flattened out to remove most of the wrinkles. If the skin is fairly pliable, the operator should attempt to place it over one of his own fingers and try several prints. If the prints secured are not suitable, the piece of skin (exert care to avoid breaking or splitting) is flattened out between two pieces of glass and photographed (figs. 397, 398, 399).

[Illustration: 397. Inked print made from the finger of deceased before treatment.]

[Illustration: 398. Inked print made after skin was removed and treated.]

[Illustration: 399. Photograph of ridge detail of skin flattened between two pieces of glass.]

The foregoing outline covers the procedures followed in cases involving decomposition in which the outer skin is still present. In many instances of decomposition the outer skin has been destroyed or is in such a condition as to be of no value. It must be emphasized again that the second layer of skin possesses the same ridge detail as the outer layer and this, though finer and less pronounced, is just as effective for identification purposes.

If, from examination, it is apparent that the outer layer of skin is missing and the second layer is intact, the finger should be cleansed, dried, inked, and printed in the usual manner. If the specimen is wrinkled but pliable it may be possible to inject tissue builder, as previously mentioned, to round out the finger, then ink and print it.

Occasionally, some of the outer skin is still attached but is of no value. This skin should be removed by carefully picking or prying it off with a scalpel in such a manner as not to destroy or injure the ridge detail of the second layer. After the outer fragments have been removed, the second layer is cleaned, inked, and printed. In the event the resultant impressions are not suitable for classification and identification purposes, the most likely reason for it is that the ridge detail is too fine to print even though there are few if any wrinkles in it. If this is the case, the finger should be cut off at the second joint and photographed. Should wrinkles which cannot be removed by injecting tissue builder, and which also preclude the taking of suitable photographs, be present, the pattern area is cut off with a scalpel from the first joint to the tip. The flesh is then cut and scraped out as *previously described, until the specimen is thin enough to flatten out between two pieces of glass which may be held together by scotch tape. The skin is then photographed.*

Occasionally, even after the flattening process it will be noted in the ground glass of the camera that the skin may be seen plainly but the ridge detail is very poor. This difficulty may be due to the poor contrast of the ridges and furrows when using direct lighting. If so, it can be overcome by scraping the skin to transparency and then photographing it by transmitted light (i.e., passing light through the skin). Sometimes, due to the condition of the skin, even though it is tissue thin, it will not be transparent. This can be overcome by soaking the skin in xylene for a few minutes and then photographing it by transmitted light while it is still impregnated with the xylene. If the substance dries too fast to permit proper photographing, the skin should be photographed while immersed in the xylene. (See subtopic of this chapter pertaining to "general photography.") Of course, after the skin has been photographed the negative should be printed to give a reverse position so that the print will be comparable with inked impressions on fingerprint cards.

Desiccation and charring

The problem confronting the fingerprint examiner in treating fingers which are desiccated or dried and shriveled is that of distending and softening the skin. Desiccated fingers are generally found to have the outer layer of skin intact and the ridge detail fairly clear. However, due to the shrinking, numerous wrinkles will be present, and as the drying process continues the skin and flesh harden until the fingers become almost as hard as stone.

It is sometimes possible to distend or swell the flesh by utilizing a 1- to 3-percent solution of sodium hydroxide or potassium hydroxide, sometimes referred to as caustic potash. As a matter of caution, this process should be tried with one finger before using it for the remaining fingers. This point of caution is made because of the reaction of the potassium or sodium hydroxide, which is actually one of destruction. While absorption and swelling of the flesh occur, the disintegrating action of the fluid may result in total destruction of the flesh.

The finger to be distended is cut from the hand at the second joint and placed in the hydroxide. When it has resumed its normal size by the absorption of the solution, it is inked and printed. There is no set time for this process. The procedure may require a few hours or as much as several days until suitable results are obtained.

After the finger has been in the solution for about 30 minutes, it should be removed and examined in order to note the extent of the swelling and the reaction of the flesh to the solution. If no material change is noted, the finger is returned to the solution. A close watch is maintained and the finger is examined from time to time.

The solution may cause thin layers of skin to peel from the finger. Should this occur, the loose skin is carefully scraped off and the finger rinsed in water for a few minutes. It is then returned to the hydroxide for continuation of the process.

If, during the course of an inspection, it is seen that the flesh is becoming too soft, the finger should be placed in a 1- to 3-percent solution of formaldehyde or alcohol for several minutes in order to harden it.

If, after several hours in the hydroxide, the finger has not reached its normal size, it should be placed in water for an hour or two. This has a tendency to hasten the swelling. When the finger is removed, it will be noted that a film has coated the surface. This coating is carefully scraped off and the finger is replaced in the hydroxide solution for an hour or so, again scraped if coated, soaked in clean water, etc. This process of alternating from solution to water, scraping, and replacing in hydroxide is continued until desirable results are obtained. The finger is then inked and printed.

The above process will so saturate the finger with solution that it may be too wet to print properly. Accordingly, the finger may be dipped into acetone for several seconds, removed, and be permitted to dry, after which it is inked and printed.

The complete process may take from several hours to as much as 10 days to secure suitable results. If the final results of the above procedure are satisfactory with the one finger being tested, the remaining fingers are given the same treatment. Care must be taken to identify each finger properly as to right index, right middle, etc., to avoid any mixup.

In the event that the reaction of the solution on the first finger treated is not satisfactory and the operator feels that it would be futile to continue the process, the finger should be removed from the solution immediately, washed carefully in water, and placed in formaldehyde to harden sufficiently for it to be handled without causing injury to the ridges. The pattern area is cut off in such a manner that sufficient surrounding surface permits the skin to be trimmed. Then from the cut side the skin is carefully scraped and cut to remove the excess flesh. While the cutting and scraping are being done, from time to time the skin should be soaked in xylene and massaged for purposes of softening to remove wrinkles. When the skin is thin enough and sufficiently pliable, the operator places the skin on his own finger, inks and prints it in the usual manner.

If the results are satisfactory, the same procedure is followed with the remaining fingers. In the event the resultant inked prints are not suitable, the skin should be scraped until it is sufficiently thin to be flattened between two pieces of glass and photographed.

Here again it is pointed out that should there be a poor contrast between the ridges and furrows when using direct lighting, the skin is scraped as thin as possible without tearing and it is then photographed by transmitted light.

There are also included, as cases of desiccation, bodies which have been burned or subjected to severe heat. Often there are cases where the skin has become loose but is hard and crisp, or where the finger has been severely burned and is reduced almost to carbon, yet is firm. In these instances the ridge detail usually has not been destroyed.

When a body which has been severely burned is located, the problems of identification should be anticipated. Accordingly, before the body is removed, a careful examination of the fingers should be made in order to determine if the removal would, in any way, cause damage to the fingers. Should it be felt that because of the condition of the body removal would cause injury to the ridge detail, securing of fingerprints at the scene, or possibly the cutting off of the hands or fingers to avoid destruction of the skin, should be considered. An examination of the fingers may disclose that the outer skin is hardened and is partially loosened from the flesh. It is sometimes possible, by twisting back and forth, to remove this outer skin intact. If this is done, the operator may place the skin on his own finger, ink and print in the usual way.

If the skin is intact on the finger and is not wrinkled, of course there is no problem and the usual method is employed to secure impressions.

Should wrinkles be present and the skin pliable, tissue builder is injected into the bulbs, which are then inked and printed.

In the event the wrinkles cannot be removed in this fashion, the pattern area is cut off and the excess flesh scraped out as before. While the scraping and cutting are being accomplished, the skin should be soaked and massaged in xylene to soften. The skin is then placed on the operator's finger, inked and printed. Should prints made in this manner be unsatisfactory, the next recourse is photography.

In some instances the fingers of burned bodies will be charred. Such cases require very careful handling as there is a probability of destroying or disturbing the ridge detail through mistreatment. In these instances the procedure is determined by the degree of charring. In extreme cases the only method of recording is by photographing, using side lighting to secure the proper contrast of ridges and depressions. Obviously, no attempt should be made to ink and roll as the pressure necessary to secure the prints would cause the skin to crumble.

In instances where the charring has not reached the extreme stage the procedures previously set forth should be applied; that is, treatment of the skin by cleaning, softening, inking and printing, or, finally, by photographing (fig. 400).

[Illustration: 400. Photograph of charts used in actual case to establish the identity of a charred body, victim of murder. Chart A shows skin removed from one of the fingers treated and photographed. Chart B shows an inked impression of the same finger during victim's life.]

Water-soaked fingers

The third and final type of case which may confront the identification officer concerns the problem of maceration, that is, long immersion of the fingers in water.

One of the cardinal rules for securing legible impressions is that the fingers must be dry. Accordingly, in these cases it becomes a matter of drying the fingers in addition to contending with other difficulties. Usually the skin on the fingers absorbs water, swells and loosens from the flesh within a few hours after immersion.

If an examination discloses the skin to be water-soaked, wrinkled and pliable, but intact, the first step is to cleanse the skin carefully as previously described. Next, wipe the fingertip with alcohol, benzine or acetone, waiting a few seconds for it to dry. The skin is pulled or drawn tight across the pattern area so that a large wrinkle is formed on the back of the finger, then the bulb is inked and printed.

If the skin is broken and hanging loose, but its pattern area is intact, it should be removed from the finger, cleansed and placed in alcohol or benzine (not acetone) for about a minute, then stretched carefully over the operator's finger so as to remove any wrinkles. It may then be printed.

Sometimes the skin is intact on the finger but so wrinkled and hard that it is not possible to draw it tight for inking. In this case it may be advisable to inject tissue builder to round out the bulbs for inking and printing. Should this fail, the ridge detail is photographed on the finger; or the skin is cut off, flattened between two pieces of glass and then photographed. Here, again, it must be pointed out that when the ridge detail does not show on the surface of the outer skin the underside should be examined, for many times the detail can be seen clearly. Should this be true, of course, the underside is photographed.

In cases where it is noted that the outer skin is gone and the finger is not saturated with water, it is possible to dry the surface sufficiently for inking and printing purposes by rolling the finger on a blotter. If this fails, the finger is wiped off with a piece of cloth which has been saturated with alcohol, benzine or acetone, after which it may be inked and printed.

Drying the fingers

In many instances it will be found that the outer skin is gone and the fingers themselves are saturated with water. A quick method of drying out the fingers is to place them in full strength acetone for approximately 30 minutes. The fingers are then placed in xylene for about an hour or until the xylene has overcome the reaction of the acetone. After removal from the xylene the fingers should be placed on a blotter until the surface of the fingers appears dry. They are then ready to be inked and printed.

It will be noted in this procedure that when the fingers are removed from the acetone they dry and harden in a matter of seconds. The purpose of the xylene is to resoften the fingers. After this treatment, should the resulting inked impressions be unsuitable for classification purposes, the ridge detail should be photographed.

X-ray photography

The use of X-ray photographs (radiography) has been advocated by some for purposes of recording the ridge details in decomposed, desiccated, or macerated cases. Briefly, the procedure involves the covering of the fingers with heavy salts such as bismuth or lead carbonate, in a thin, even film over the pattern area and then, by the use of the X-ray, reproducing the ridge detail. This procedure necessitates the use of X-ray equipment and a technician skilled in making radiographs. It is, therefore, an expensive operation. The results of the radiograph in no way compensate for the expense, time, and skill required inasmuch as in those cases where many wrinkles and creases appear in the fingers, especially desiccated specimens, the results have been very poor. In instances where there are no wrinkles or only a few, and where the creases are not too deep, the ridge detail is reproduced very well in the radiograph. In these cases, however, it is usually possible to secure impressions by inking and rolling in the regular way or, should this fail, ordinary photography will certainly give satisfactory results. For economical and practical purposes the use of the X-ray is not recommended.

General photography

In the foregoing instances in which it has been impossible to obtain suitable inked impressions it will be noted that the last resort has always been photography. In all probability in advanced cases of decomposition, desiccation, and maceration it may not be possible to secure inked impressions which can be properly classified. Hence, it will be necessary to photograph the ridge detail. Accordingly, there are outlined below several methods of photographing the ridges which have been used with success.

In photographing the ridge detail on fingers it has been determined to be most practicable to photograph the finger natural, or 1/1, size inasmuch as comparisons will usually be made with inked impressions which are natural size. Any camera built or adjusted to taking 1/1 size pictures, and with which the lighting may be arranged to best advantage, may be used.

There is a wide choice of film which can be used for this purpose. The so-called soft films are all good for photographing ridge detail on fingers. Process film is not recommended inasmuch as the film presents too much of a contrast. Consequently, if it is used, some of the ridge detail will be lost, especially if wrinkles are present in the skin.

Lighting is accomplished by the use of gooseneck lamps, floodlights, or a spotlight. If a fingerprint camera is used, its lights may be sufficient.

The manner of lighting may be by direct light, side light, transmitted light or reflected light, depending upon the prevailing condition of the finger or skin.

Direct light is used in those cases in which the ridge detail is fairly clear and there are no wrinkles present; or, if wrinkles are present, they are not deep enough to interfere with photographing the ridges.

Side lighting is used when there are no wrinkles of any consequence and the ridge detail is clear but because of discoloration the ridges are not readily seen in the ground glass as there is lack of contrast between ridges and depressions. Accordingly, the lights, instead of being focused directly on the skin or finger, are placed to the side of the object so that the light is directed across the skin or finger, thus highlighting the ridges and shading the depressions.

In side lighting, two lights may be used. Better results are often obtained, however, by using only one light, such as a spotlight, the beam of which can be controlled to best advantage.

Transmitted light is used in cases in which the skin has peeled off or in which the dermis has been removed, cut, and scraped thin so that light will go through. The prepared skin is placed between two pieces of glass pressed together in order to flatten the skin or dermis and remove creases. By trimming some of the surplus skin or dermis, especially at the top, it may be more easily flattened. After the glass is properly mounted in front of the camera, the lights are placed behind it and light is directed through the skin. The ridge detail is brought into focus on the ground glass. Before the picture is actually taken it is suggested that the ground glass be checked by first using one light and then two lights to see which is more effective.

There will be instances in which the second layer of skin, cut and scraped thin enough to flatten out, fails when dry to have a sufficient contrast between ridges and depressions for purposes of photographing. The same piece of skin when soaked in xylene will show a marked contrast, which it loses on drying. This difficulty is overcome by photographing the skin while in solution, which can be done by placing the skin in a test tube or a small bottle of a size to keep the skin upright and the ridges toward the camera. The test tube or bottle is then filled with xylene.

CHAPTER X 129

If the skin is sufficiently thin, transmitted light may be used. Should it be found, however, that transmitted light is not effective, then direct light may be tried and the results checked in the ground glass (fig. 401).

When photographing a small curved surface such as a test tube, direct lighting will usually create a high light. If the high light as shown in the ground glass is over the ridge detail on the skin, a poor photograph will result. If the high light cannot be removed by rearranging the lights, then reflected light should be tried.

In order to effect reflected light a large piece of white paper, cardboard, or similar material is used. A hole is cut in the center of the paper or cardboard. This must be big enough for the camera lens to protrude through. The ends of the paper or board are curved toward the skin or finger to be photographed. The lamps which are to be used are placed facing the curved paper or cardboard in such fashion that the light will strike the paper or board and be reflected by the curved surface to the object.

The lamps should be close enough to the paper or board to give the maximum light. Care should be exercised, however, not to place them too close, because of the fire hazard.

Any arrangement of lamps and reflectors giving a similar effect as the above should prove suitable.

Fingers or skin which have a mottled, reddish-brown color because of decomposition, exposure to severe heat, or diffusion with blood present a problem of lack of contrast between ridges and depressions for photographic purposes. This lack of contrast can be overcome to a large extent by the use of a yellow or light red filter. Sometimes, in those cases where the discoloration is due to the diffusion of blood throughout the tissues, the blood can be washed out by saturating and rinsing the specimen in a 10- to 20-percent solution of citric acid. If, of course, the blood is not removed satisfactorily, the photographing should be done with the filter.

[Illustration: 401]

CHAPTER X   130

As previously stated, the fingerprint camera can be readily adapted to the use of photographing fingers or skin specimens for ridge detail. Sometimes it is possible to photograph the skin or finger in the same manner as one does a latent print. There will be instances, however, in which the standard use of the fingerprint camera will not be possible or effective, such as for side light, reflected light, and sometimes transmitted light, or instances in which it is not possible to get the finger or skin flush with the opening of the camera. In these instances the lights of the camera are not used, so the batteries should be removed and gooseneck lamps or other suitable lighting equipment and ground glass utilized when the finger or skin is prepared for photographing (fig. 402).

The camera is opened either at the point where the lights are housed or at the lens point, whichever is most effective. Then, opening the shutter, the operator moves the camera either toward or away from the finger or skin to the point where the ridge detail is sharpest in the ground glass. The camera is held firmly, the ground glass is removed, the film is inserted and the photograph taken.

[Illustration: 402]

With respect to exposure time, it is possible only to generalize and point out that each case will have its own individual aspects. Controlling features for consideration will be the type of film, the type and size of lights, the method of lighting (direct, side, transmitted or reflected) and also whether or not filters are used. Accordingly, there may be a wide variation of exposure time in different cases.

The best approach for arriving at the proper exposure time is merely to make a test exposure, develop the film, and from an examination determine if it is underexposed or overexposed. Time the next exposure accordingly, until satisfactory results are obtained.

As has been mentioned previously, when photographing the ridges on fingers or skin, the ridge detail will be in reverse position, the opposite from an inked impression made from the same skin or finger. (This is true except in those cases in which the underside of the epidermis is photographed.) Accordingly, when the negative is printed, it should be

CHAPTER X   131

printed gloss side to sensitive side of paper to give the position comparable to an inked print made from the same skin or finger. In order to avoid error or confusion a notation should be made on the photograph of each finger, or, if they are cut and mounted on a fingerprint card, point out that the position has been reversed and that the prints are in their correct position for classifying and searching. Otherwise, it is possible that the right hand may be mistaken for the left hand and vice versa.

If the underside of the epidermis or outer skin is being photographed, the negative should be printed in the normal manner, that is, emulsion side of negative to sensitized side of paper. Here, reversal of position is not necessary for when the ridge detail is viewed from the underside it appears to be in the same position as the inked impression normally is reflected on a fingerprint card.

Care should be taken to see that each photograph is labeled correctly to indicate the finger it represents, such as right thumb, right index, right ring, etc. It is imperative that no error occurs in such labeling, inasmuch as it is highly probable that the resultant classification would be incorrect and failure to make an identification might very easily follow.

Deceased infants

The foregoing has dealt with the securing of fingerprints of unknown deceased persons for identification purposes. The basis for such action is the presumptive possibility that the unknowns had been fingerprinted previously and through this medium might be identified.

Another type of problem, however, arises with the finding of a deceased infant. It can be safely assumed that the possibility of there being in existence a set of known fingerprints of the infant is extremely remote. Nevertheless, in view of the fact that many hospitals throughout the country, as part of the general routine of recording a birth, secure the infant's footprints, it would follow that there could be a remote possibility of identifying the deceased infant through its footprints. The foregoing principles and procedures would then apply in securing the foot impression of a deceased infant. It is fully realized that in practically all cases involving the finding of an abandoned infant corpse the infant is probably illegitimate issue and delivery did not occur in a hospital, but there have been instances where such was not the case.

The importance of securing footprints of deceased infants killed in a common disaster cannot be overemphasized. Such disasters may involve the death of infants of lawful issue, and in many instances there are hospital footprint records available which may prove of value as a positive means of identification.

Technical consideration

The methods described are intended to record, either by printing with ink or by photographing as legibly as possible, the ridge details of the tips of the fingers of unknown dead for identification purposes. The securing of the impressions enables the fingerprint examiner to classify and search them through a file. This "search," of course, means merely to make a comparison of the deceased's prints with the prints of known individuals.

It is well to bear in mind the fact that the dermis or epidermis may have undergone certain physical changes and that in order for the fingerprint examiner to make a proper comparison he must know the changes which can and do occur. Otherwise, he may fail to make an identification (fig. 403).

[Illustration: 403. Epidermis or outer layer of skin commencing to peel from dermis or second layer of skin, result of decomposition.]

Consider first the epidermis or outer layer of skin in cases of maceration (the skin is water soaked). There may be considerable swelling. The ridges become broader and are more distinct. An inked impression in such an instance may show a pattern larger in area than a print made from the same finger when the person was alive. Also, if the skin is on the finger but is loose, inking and rolling could distort the impression so that some of the ridge formations would seem to be in a different alignment from corresponding details in a print made during life. When decomposition commences, what are really solid ridges may be broken, giving rise to the possibility that there appear to be more characteristics than there actually are (figs. 404 and 405).

[Illustration: 404. Inked fingerprint made during life.]

[Illustration: 405. Inked impression of same finger of deceased showing effect of decomposition.]

The existence of wrinkles may also cause the impression to acquire an appearance of dissimilarity when compared with the original inked print.

With respect to cases of desiccation, there will probably be shrinkage, hence, the impressions made may appear smaller than in life and the ridges will be finer. In cases in which the epidermis has been lost and there remains only the dermis or second layer, there will usually be shrinkage with the same results. Here also, wrinkles, if present, may cause a difference in appearance from the normal print.

[Illustration: 406]

In addition to shrinkage and wrinkles in cases involving the second layer of skin, there is a radical change in the appearance of the ridges themselves. The second or dermal layer of skin is composed of what are called dermal papillae which have the appearance of minute blunt pegs or nipples. The dermal papillae are arranged in double rows (fig. 406). Each double row lies deep in a ridge of the surface or epidermal layer and presents the same variations of ridge characteristics as are on the outer layer of skin except that they are double. Accordingly, when the second layer of skin is printed or photographed, the ridge detail will appear in double. That is, the ridges will appear as though they were split. This may well confuse the fingerprint examiner in that what may be a loop having 10 ridge counts may appear to be a loop having 20 ridge counts when the impression is made from the second or dermal layer of skin. These double rows of ridges are finer and not as sharp as the detail on the outer skin, which adds to the difficulty of arriving at a correct classification and making a proper comparison.

FBI aid

The above techniques and procedures have been dealt with upon the basis that the law enforcement officers would, when a corpse has been found, attempt to secure a set of finger impressions in an effort to identify the unknown dead. If, however, the officer feels that the job is too difficult or is beyond his scope, consideration should be given to cutting off the hands or fingers of the deceased and forwarding them to the Identification Division of the FBI for processing. If this course is decided upon, it is reiterated that local statutes governing the cutting of the dead must be complied with and proper authorization must be secured.

[Illustration: 407]

In order to facilitate the transmission of such specimens to the FBI the following suggestions are made:

First, it is deemed most desirable, when possible, to have both of the hands, severed at the wrist, forwarded in their entirety (fig. 407). It is desired that the hands, rather than each separate finger, be sent inasmuch as it eliminates the possibility of getting the fingers mixed up or incorrectly labeled. If, however, it is not possible to send the hands for some reason, then, of course, the fingers should be cut off and forwarded. In cutting, the fingers should be cut off at the palm beginning with the right thumb, then the right index, ring, etc., just as though they were to be printed. As soon as each finger is cut off it should be placed in an individual container, such as a small glass jar, and immediately marked as to which particular finger it is.

In the event that the hands or fingers of more than one dead are being transmitted, it is absolutely necessary that not only the fingers be properly labeled but that each body also be given an identifying number or symbol which must be indicated on the hands or fingers cut from that body as well, in order to avoid the embarrassing situation of identifying the hands and not knowing from which body they were cut.

In shipping, the hands, fingers, or skins may be placed in preserving solutions such as 5-percent solution of formaldehyde, 5-percent solution of alcohol, or embalming fluid. When hands or fingers are desiccated (dried out), however, it is most desirable that they be placed in airtight containers and sent without any preservative. If glass containers are used, the specimens should be packed in such a manner as to avoid breakage. Dry ice is a suitable preservative for transmitting such specimens but it should not be used when shipping will take more than 24 hours.

In making up a package using dry ice, the hands or fingers, properly tagged, should be placed in cellophane or paper bags. A material such as sawdust, shavings or similar packing which acts as an insulation is placed around the specimens. A sufficient amount of dry ice is then placed in the package which is then packed tight with more sawdust or shavings. The dry ice should not be in direct contact with the cellophane or paper bags which contain the hands or fingers.

A letter covering transmittal of the specimens should be prepared in duplicate. It should, of course, indicate the sender. The names of any probable victims, sex, race and approximate age of the deceased should, if such information is available, be secured from the coroner or medical examiner and be included in the letter. A copy of the letter should be placed in the package. The original should be mailed separately. Both letter and package should be addressed as follows:

DIRECTOR FEDERAL BUREAU OF INVESTIGATION U.S. DEPARTMENT OF JUSTICE WASHINGTON 25, D.C.

Attention: Identification Division--Latent Fingerprint Section.

If the package contains glass jars it should be marked "Fragile" to insure careful handling in transit.

The package should be sent railway express, prepaid, or, where there is need for speed, by air express, prepaid. When they are received by the Identification Division, the specimens will undergo various treatments which may necessitate further cutting, scraping, etc. In all cases, regardless of condition, the specimens will be returned after examination.

All of the foregoing matter has dealt with instances in which it has been assumed that all ten fingers are available, or a sufficient number of the fingers of a deceased have been secured and impressions suitable for searching through the fingerprint files of the FBI have been recorded.

There will be cases, however, where only a few, or possibly only one, of the fingers has sufficient ridge detail for identification. In such instances a search through the FBI files would be impractical. This, however, does not preclude the possibility of making a positive identification by the use of the one finger. Though a search through the file is not possible, a comparison can be made with the fingerprints of individuals who it is thought the deceased may be or, in some instances, with the fingerprints of missing persons.

In this connection, where one or only a few fingers are forwarded to the FBI, the names of all possible victims should also be submitted. The fingerprints of those individuals, if available, will then be taken out of file and compared with the ridge detail on the finger of the deceased in an endeavor to establish a positive identification. Many such identifications have been effected.

In conjunction with the usual services afforded authorized law enforcement agencies, the services of an FBI fingerprint expert are also made available in those cases where expert testimony is necessary to establish the identity of the deceased through fingerprints, providing, of course, such an identification has been made.

Extreme caution should be exercised in the case of the chemicals previously mentioned in this article. Acetone, alcohol, benzine, and xylene are highly inflammable and should neither be used near open flames nor while the operator is smoking. The fumes given off by acetone, benzine, xylene, and formaldehyde are toxic and may cause sickness. They should be used in a well-ventilated room only. It is also suggested that the fingerprint examiner wear rubber gloves when using acetone, benzine, xylene, formaldehyde, potassium hydroxide, or sodium hydroxide. These chemicals will cause the skin to peel. Strong concentrations may cause burns.

In conclusion, it is pointed out that the procedures and techniques which have been described are those currently being used by the *fingerprint experts of the FBI. These methods are fast and the results have been most satisfactory. This Bureau does not claim, however, that satisfactory results cannot be achieved through variations thereof or different methods.*

CHAPTER XI

Establishment of a Local Fingerprint Identification Bureau

For the information and assistance of officials who desire to establish a local fingerprint identification bureau, the following suggestions are being made to indicate the principal materials necessary to equip such a bureau:

Fingerprinting equipment

For the purpose of taking fingerprints there should be a stand with a clamp for holding the fingerprint cards steady. This latter item is necessary to prevent smudging the prints. A tube of printer's ink is used. The ink is applied by a roller to a glass plate upon which the fingers are inked before being rolled on the cards. The complete equipment for the above process may be secured from a number of commercial sources or it can be made. Figure 408 depicts an inking stand.

Fingerprint files

It is suggested that the fingerprint card be white, light cardboard, 8 by 8 inches, slightly glazed. This size is convenient, as it allows all the space necessary for recording the classification of the prints and general descriptive information concerning the individual. In the event the new bureau desires to contribute copies of its fingerprints to the Federal Bureau of Investigation, the latter will, upon request, gladly furnish fingerprint cards for the purpose together with envelopes and instructions on how to take fingerprints. It is suggested that the new bureau design its cards similar to those furnished by the Federal Bureau of Investigation, as these have been designed after special study and have been found to be satisfactory over a long period of time. Figures 409 and 410 show the fingerprint side and reverse side of the criminal fingerprint card used by the Federal Bureau of Investigation.

In classifying and comparing fingerprints it is necessary to use a magnifying or fingerprint glass. Such instruments can be obtained from

CHAPTER XI

various commercial sources. Figure 411 shows the type of magnifying glass used by the Federal Bureau of Investigation.

[Illustration: 408. Diagram of a FINGERPRINT INKING STAND]

The fingerprint cards should be filed according to fingerprint classification sequence in cabinets, preferably steel. It is further suggested that the cabinets be three drawers high, with each drawer divided into three rows for filing. Such cabinets or similar ones can be obtained from various commercial sources. Figure 412 shows the type of fingerprint cabinet used in the Federal Bureau of Investigation.

In order to facilitate the location of classification groups, it is suggested that guide cards be placed in the rows of fingerprint cards at frequent intervals. These guide cards should be slightly longer and heavier than the fingerprint cards and should have small tabs on the top to hold classification identifying symbols. Figure 413 shows the type of guide card used by the Federal Bureau of Investigation.

[Illustration: 409]

LEAVE THIS SPACE BLANK |TYPE OR PRINT |SEX |RACE | | |
|LAST NAME FIRST NAME MIDDLE NAME|--------|----- | |HT. |WT. |
|(Inches)| | | |------------------------------------|--------|----- |CONTRIBUTOR AND |ALIASES |HAIR |EYES |ADDRESS | | | | | |--------------
------------------------| | |DATE OF BIRTH SIGNATURE OF PERSON | | | FINGERPRINTED | | |-------------- | | |PLACE OF BIRTH | | |
|-- ------------------------|YOUR NUMBER |LEAVE THIS SPACE BLANK SCARS AND |AMPUTATION | | MARKS | |------------------|CLASS | |PLACE FBI NUMBER |
------------------------- -------------------------|HERE | SIGNATURE OF DATE |------------------|REF. OFFICIAL TAKING | CHECK IF NO |
------------------------- FINGERPRINTS ||/ REPLY | | IS DESIRED |
--- 1. RIGHT THUMB|2. RIGHT INDEX|3. RIGHT |4. RIGHT RING |5. RIGHT | | MIDDLE | | LITTLE | | | |
--- | | | |
--- 1. LEFT THUMB |2. LEFT INDEX |3. LEFT MIDDLE|4. LEFT RING |5. LEFT

CHAPTER XI

LITTLE | | | | | | | |
-- | | | |
-- LEFT FOUR FINGERS TAKEN |LEFT |RIGHT |RIGHT FOUR FINGERS TAKEN SIMULTANEOUSLY |THUMB |THUMB |SIMULTANEOUSLY | | | | | | | |
| | | |

A practice which has been of the utmost benefit in the Federal Bureau of Investigation is as follows: When a fingerprint card is taken out of its regular file for any reason, a substitute card is put in its place, to remain until the return of the card. This substitute card, or "charge-out" card, is of a different color from the fingerprint card and slightly longer. On it are recorded the name, the classification formula, and peculiar characteristics, such as scars and peculiar pattern formations, appearing on the original card. By indicating the date and reason for charging out the original card, the Bureau is able to keep an accurate check on the whereabouts of all prints at all times. It is suggested that the local bureaus adopt a practice of this kind whenever a fingerprint card is drawn from the files and it is known that it may be out for a period of time longer than the remainder of the day on which it is drawn. Figure 414 shows the type of charge-out card used in the FBI.

[Illustration: 410]

FEDERAL BUREAU OF INVESTIGATIONS, UNITED STATES DEPARTMENT OF JUSTICE WASHINGTON, D.C.
-- CURRENT ARREST OR RECEIPT
-- DATE ARRESTED | CHARGE OR OFFENCE |DISPOSITION OF SENTENCE OR RECEIVED |(If code citation is used |(List final disposition only. |it should be accompanied by|If not now available |charge) |submit later on FBI Form R-54 | |for completion of record.) | | |
| ---------------|--------------------------| OCCUPATION |RESIDENCE OF PERSON | |FINGERPRINTED | | | --| If COLLECT wire reply or COLLECT telephone | reply is desired, indicate here | | |\ Wire reply |/ Telephone reply | | FOR INSTITUTIONS USE ONLY ---------------- | Telephone number | Sentence expires_____ --|--------------------------

CHAPTER XI

| INSTRUCTIONS |1. FORWARD ARREST CARDS TO FBI | IMMEDIATELY AFTER | FINGERPRINTING FOR MOST | EFFECTIVE SERVICE. | |2. TYPE or PRINT all | information. | |3. Note amputations in proper | finger squares. | Please Paste Photograph in This Space |4. REPLY WILL QUOTE ONLY | NUMBER APPEARING IN THE Since photograph may become detached | BLOCK MARKED "CONTRIBUTOR'S indicate name, FBI number, and arrest | NO." number on reverse side whether attached to | fingerprint card or submitted later. |5. Indicate any additional | copies for other agencies | in space below--include | their complete mailing | address.
--------------------------------------|------------------------------ | SEND COPY TO: | | | | --
FD-249

Each fingerprint card handled by the bureau should receive a fingerprint number and these numbers can be assigned in consecutive order.

As the bureau increases in size, it will be found a source of much convenience to have fingerprints of males and females kept in separate files.

Name files

There will be times when it may be necessary to locate an individual's fingerprints when no current fingerprints are available, but the name with a police number or the classification is known. In order to facilitate work of this nature, as well as to keep a complete record and check on each set of fingerprints, it is necessary that the files be indexed in a manner similar to that in which books in a library are indexed.

[Illustration: 411]

[Illustration: 412]

In this connection, for each fingerprint card there is prepared an index card. On this the name of the individual is placed, with all known aliases, the fingerprint classification formula, the police or arrest number, the date of arrest or other action. It is desirable, also, to have on this card such general information as age, height, weight and race. Figure 415 shows the front of a suggested type of 3- by 5-inch index card.

[Illustration: 413]

[Illustration: 414]

```
/ / / / / / / / / / / / / / / / / / / |FBI NUMBER |Followed
+--------+--------+----------+----------+-------+-------------+ |ASSEMBLY|
POST G | N-IDENTS | DOC. LAB | MISC. | |
+--------+--------+----------+----------+-------+ | |Searched Thru|
+-------------+
```

.... Classification.................... MASTER PRINT
NAME

.... Reference........................ CURRENT PRINT
NAME

```
+--------------+--------------+---------------+-------------+-------------+ |1. Right
Thumb|2. Right Index|3. Right Middle|4. Right Ring|5. Right | | | | |
|Little | / / / / / / / / / / / /
+--------------+--------------+---------------+-------------+-------------+ |6. Left
Thumb |7. Left Index |8. Left Middle |9. Left Ring |10. Left | | | | |Little |
/ / / / / / / / / / / /
+--------------+--------------+---------------+-------------+-------------+ TYPE OF
```
CURRENT PRINT |Number and Initials of Criminal | Non-Criminal |Employee Charging Out: | |
Date.................|Date...................|............................ 16-58188-1 U.S. GOVERNMENT PRINTING OFFICE

Figure 416 shows the reverse side of the 3- by 5-inch index card. These are filed alphabetically in special cabinets. An index card also should be made for every alias which an individual has used. Figure 417 shows an electrically operated file cabinet in which the index cards are filed. It is suggested that the alias cards be of a different color from the one bearing the correct name, known as the "Master." Each alias card also should have typed on it the correct name of the individual, for

CHAPTER XI

purposes of reference and cross-checking. For convenience and accuracy these files, as in the fingerprint files, should also have suitable guide cards.

[Illustration: 415]

```
/ /
_____/_____/_____
LAST NAME FIRST NAME MIDDLE NAME | F.P. CLASS | IDENT NO.
/ / /
_____/_____/_____/_____
ALIAS RACE SEX AGE


_____
ADDRESS | | | | | | | | | |
_____/_____/_____/_____/_____/_____
HEIGHT WEIGHT EYES HAIR COMPLEXION OCCUPATION |
_____/_____
DATE AND PLACE OF BIRTH SCARS AND MARKS
_____
(SEE OTHER SIDE FOR ARREST RECORD)
```

[Illustration: 416]

```
DATE | NUMBER | CHARGE | DISPOSITION
____/_____/_____/_____/_____/_____/_____
/ / /
```

It is advisable to make use of charge-out cards when original index cards are drawn from the files. Figure 418 shows a charge-out card.

To supplant the 3- by 5-inch index cards mentioned above, many law enforcement agencies have found it desirable to use a separate sheet, sometimes referred to as a "History Sheet" or "Information Sheet," containing the complete case history of the subject involved. These separate sheets can be filed by fingerprint number sequence and contain not only the data such as the known aliases, the fingerprint classification formula, the arrest number, and other essential items which are set out on the 3- by 5-inch cards as heretofore described, but also contain a concise summary of the subject's arrest history, particularly with reference to his criminal activities in the particular city. They may also contain a summarized case history with respect to each arrest or commitment, including such items as the date and place of arrest, complete home address, relatives, the essential facts concerning the prosecution of charges, and the ultimate disposition.

[Illustration: 417]

Jacket folder file

When an identification bureau receives prints of individuals on whom it already has prints, it is not practical to keep more than one set of prints per person in what may be called the active fingerprint file. In these instances the better print should be designated a "Master" print by having the word "Master" stamped thereon. It should be given a number, to be known as the master number, which number should also be placed on all other sets of prints which may be found to be identical with the "Master" print. The "Master" print is placed in the active files. The extra prints are placed together in a heavy folder with their master number stamped thereon. This jacket folder is then filed in a separate cabinet. Also, if copies of all information regarding an individual, photographs, and FBI transcript of record are kept in this folder, his complete record will always be assembled in an easily accessible unit. The "Master" number should also be placed on the index card and all the alias cards of the individual. Also, each new alias and arrest number should be placed on the original index card. The additional records are kept in folders which are arranged in numerical order, beginning with Nos. 1, 2, 3, and so on.

[Illustration: 418]

1-154 |-----------| | GPO: 1962 OF--663475 | | | |-----------|
|------------------------ | | | POSTING |-----------| |------------------------ | | | | |
|DOB -----------------------|-----------|-----------|------------------------ |
ASSEMBLY | MISC. | DESCRIPTION

NAME_____

CHAPTER XI   145

NAME_____

(ARREST NUMBER) (DEPARTMENT) (CITY) (STATE)

A further suggestion in connection with the maintenance of this folder file, besides the use of a separate "Master" numbering system, is the use of the arrest fingerprint number. As indicated previously, each person arrested and fingerprinted is assigned a number. This number appears on the fingerprint card, name-index card, and photograph. The practice of handling every new arrest fingerprint card in the bureau should include searching the fingerprint card in the fingerprint file to ascertain if the subject has a previous record. If the subject does not have a previous record, a new number should be assigned. In this connection it is noted that only one copy of the fingerprint card should be maintained in the file by fingerprint classification. To indicate the new arrest on the old index card, the date of the new arrest can be shown. Whether the bureau follows the "Master" numbering system or the "previous arrest" numbering system should make very little difference in the ultimate purpose. All extra copies of fingerprint cards, complete record sheets, photographs, and all information pertaining to the individual are filed away in the folder file. This complete record is readily accessible at all times. It will now be found that the bureau has a complete record of each individual who has an arrest record on file, with provision made for accurate cross-referencing and checking between names and fingerprints.

Dispositions

It is important to the bureau to have complete information concerning the ultimate disposition on each arrest fingerprint card. If the disposition of a charge is known at the time the person is fingerprinted, this fact should be indicated in the space provided on the fingerprint card. For example, in the case of an individual who is arrested, fingerprinted, and turned over to the county jail, this disposition can be indicated on the fingerprint card which is forwarded to the Federal Bureau of Investigation. The fingerprint card should not be held by the bureau pending final disposition of the charge.

In those cases where the disposition is pending prosecutive or court action, a separate 3- by 5-inch disposition file can be maintained. On these cards information concerning the name, fingerprint number, race, sex, charge, name of the arresting officers, and the fingerprint classification should appear. These cards are filed in a pending-disposition file. The 3- by 5-inch disposition cards are made at the time the fingerprints of the person are taken. When the final disposition is obtained it should be noted on the card. In those cases where there is only one fingerprint card in the bureau, the disposition can be noted on the name-index card or the reverse side of the bureau's fingerprint card. In those cases where there is a jacket-folder file for the individual this disposition card can then be placed in the folder.

"Disposition Sheets" (No. R-84) can be obtained from the Federal Bureau of Investigation for forwarding this information so that the files of the FBI will have complete information concerning the arrests. At the time the final disposition is obtained, these disposition sheets can be completed and forwarded to the Federal Bureau of Investigation.

Death notices

When persons whose fingerprints are on file are reported as deceased, the prints should be taken from the active file and assembled with any other prints of the person concerned. These should be plainly marked "Dead" and filed in a separate cabinet or section. All the index cards on this individual should also be marked "Dead" and filed in a separate section. These should be retained for possible future reference.

In this connection, "Death Notice" forms (No. R-88) can be obtained from the Federal Bureau of Investigation so that information concerning these deaths can be properly noted in its fingerprint file.

Record of additional arrest

It is not necessary for a bureau to send a regular fingerprint card to the Identification Division of the Federal Bureau of Investigation on individuals who have been arrested repeatedly and whose previous records are known to the local law enforcement agency. In such cases the "Record of Additional Arrest" form should be used.

Complete information must be given on this form. It is imperative that the FBI number and the finger impressions be placed on this form. The Identification Division of the Federal Bureau of Investigation will send no answer upon receipt of this form.

The form will be placed in the FBI number folder on the individual and later when a regular fingerprint card is received the arrest information from all the forms will be compiled and included on the subject's record as "supported by fingerprints."

"Record of Additional Arrest" forms (No. 1-1) can be obtained from the Federal Bureau of Investigation.

Wanted notices

All wanted notices containing fingerprints, including the wanted notices inserted in the FBI Law Enforcement Bulletin, should be filed in the fingerprint file by classification formula, and the names appearing on these wanted notices should be indexed and placed in the name files. Concerning the small wanted notices inserted in the FBI Law Enforcement Bulletin, a suggested procedure would be to paste each individual notice on a blank 8- by 8-inch white card. The wanted notices are filed by the fingerprint classification and the names indexed and placed in the name file. When an apprehension notice is received concerning the wanted notice, a proper notation should be made on the name card and the wanted notice in the fingerprint file. If these canceled wanted notices endanger the efficiency of the file, it is suggested that the name-index card and the fingerprint-wanted notice be destroyed. Should the bureau adopt this practice it is suggested that the 8- by 8-inch cards be used again for other wanted notices. In this manner it would be possible to use the blank card for eight of these notices.

CHAPTER XI 148

The Federal Bureau of Investigation will make available to law enforcement agencies a special "Wanted Notice" form (No. 1-12) in order that they can place wanted notices against the fingerprints in the files of the FBI.

Photographs

Arrangements should be made to procure a camera for taking photographs of the persons fingerprinted. This is known as a "mugging" camera and various types are on the market. It is believed that the photographs should include a front and side view of the person. In most instances a scale for indicating height can be made a part of the picture even though only the upper portion of the individual photographed is taken. Of course, if the scale is used, the person photographed should be standing even though only the upper portion of the body appears in the photograph. The necessary lights should be provided for obtaining photographs. A standard set of scales should be obtained in order that the correct weight can be ascertained.

[Illustration: 419]

The negatives and photographs can be filed by fingerprint number in a separate file. In those cases where the individual has more than one arrest all the photographs can be placed in the jacket-folder number file. The negatives, in these instances, can remain in the photograph file.

Latent fingerprints

To adequately develop the latent prints at crime scenes, it is necessary that the proper equipment be provided. This equipment includes latent fingerprint powders, brushes, lifting tape, fingerprint camera, searchlight, and scissors. All of this equipment can be obtained from commercial fingerprint supply companies. Figure 419 shows some of the equipment used by the FBI. The techniques of developing latent fingerprints and their uses are more fully explained in the following chapters.

It is believed that by following the basic ideas outlined, the officials of law enforcement agencies can be assured of best results in establishing and maintaining a small identification bureau. For further information, the Federal Bureau of Investigation will furnish to duly constituted law enforcement officials any additional data which may be of material assistance in the maintenance of such a bureau.

CHAPTER XII

Latent Impressions

Each ridge of the fingers, palms, and soles bears a row of sweat pores which in the average person constantly exude perspiration. Also, the ridges of the fingers and palms are in intermittent contact with other parts of the body, such as the hair and face, and with various objects, which may leave a film of grease or moisture on the ridges. In touching an object, the film of moisture and/or grease may be transferred to the object, thus leaving an outline of the ridges of the fingers or palm thereon. This print is called a latent impression, the word "latent" meaning hidden, that is, the print many times is not readily visible.

Latent impressions, regardless of the area of the ridges present, are of the greatest importance to the criminal investigator as identification of them may solve the crime and result in successful prosecution of the subject. Consequently, every effort should be made to preserve and identify them.

Visible prints in mediums such as blood, grease, dirt, or dust are equally important to the investigator but, strictly speaking, are not latent impressions.

A search of the crime scene should be conducted in a logical manner. Points of entry and exit should be examined, along with surfaces or objects disturbed or likely touched during the commission of the crime. The examiner should wear a pair of light cloth gloves and handle an object only insofar as is necessary and then only by edges or surfaces which are not receptive to latent impressions. A record of the exact location of a print on an object and of the object itself should be made, since these facts may be of the utmost importance in any trial resulting from the investigation. No one should handle an object other than the examiner himself.

Portable articles removed should be labeled or marked so that they may be readily identified thereafter.

The beam of a flashlight played over the surface of an object will frequently show the location of latent impressions, although this is not an infallible test for their presence.

Evidence should be examined as soon as feasible after its discovery.

Following the location of any latent prints at the scene of a crime, the prints of all persons whose presence at the place under inspection has been for legitimate purposes must be excluded from further attention. It is advisable, therefore, during the initial stages of an investigation where latent prints are found, to secure the inked prints of all members of the household, the employees, and any police or other officials who may have touched the objects on which the latent impressions were found. Inked prints taken for this purpose are referred to as elimination prints.

Due to the fragmentary nature of most latent prints it is not possible to derive a classification which makes a file search practicable. A latent impression may be identified, however, by comparison with the prints of a particular suspect.

Inked fingerprints taken for comparison with latent impressions should be as legible and as complete as possible, including the areas not essential to classification, since identifications are often made with these areas. Inked palm prints taken should likewise be complete and clear and should include impressions of the finger joints. Persons not experienced in latent print comparisons should not attempt to evaluate latent fragments, since the area necessary for an identification may be extremely small compared to that of an average inked fingerprint.

Articles which are to be transported by mail or express should be so packed that the surfaces bearing latent impressions are not in contact with other surfaces. This may be accomplished by mounting the articles on a piece of fiber board or plywood. The board should then be secured in a box so that the objects will not touch or be shaken against the sides in transit. The package should be plainly marked "Evidence," to prevent inadvertent handling on opening. Cotton or cloth should never be placed in direct contact with any surface bearing latent prints.

Any number of paper or cardboard specimens may be placed in a single protective wrapper, since contact with other surfaces does not harm latents on such objects. Lifts, negatives and photographs are readily enclosed with letters.

An explanatory letter should accompany all evidence. If it is necessary to pack the evidence separately, a copy of the letter should be placed in every package so that the recipient will know immediately the import of the contents. All items of evidence should be marked and described exactly in the accompanying letter so that they will not be confused with packing material of a similar nature, and to provide a check on what the package should contain.

In addition, the letter should include for record purposes a brief outline of the crime, i.e., type, date and place of occurrence, and names of victims and subjects. If suspects are named for comparison, sufficient descriptive data should be set out to permit location of their fingerprint records. This information, in preferential order, comprises the individual's complete name, aliases, FBI number, date of prior arrest or fingerprinting, fingerprint classification, date and place of birth, and physical description.

Evidence is preferably forwarded by registered mail or railway express, as these means provide records of dispatch and receipt.

Elimination or suspect fingerprints are best enclosed with the evidence itself, with a notation as to the type of prints forwarded.

CHAPTER XIII

Powdering and Lifting Latent Impressions

The sole purpose in "developing" a latent impression is to make it visible so that it may be preserved and compared. Various powders and chemicals are used for this purpose. When a latent print is plainly visible, it should be photographed before any effort is made to develop it.

No attempt should be made to brush or apply powder to prints in dust, obviously greasy prints, or bloody prints, as this will almost surely destroy them. Objects which have been wet or immersed in water may still bear identifiable latent impressions. Before any examination is attempted, however, the object must be dried.

Powder brushed lightly over a latent-bearing surface will cling to grease or moisture in the ridges of a latent print, making it visible against the background. Obviously, a powder should be used which will contrast with the color of the surface. Photographic contrasts should also be considered.

A gray powder and a black powder are adequate for latent print work. Many fingerprint powders of various colors and compositions are available from fingerprint supply houses but none are superior to the gray and black.

A very small amount of powder is placed on the brush for application to the surface. Once the contour of a print is visible, the brush strokes should conform to the direction of the ridges. All excess powder should be brushed from between the ridges. Too much powder and too little brushing are the chief faults of beginners.

Gray powder is used on dark-colored surfaces. It is also used on mirrors and metal surfaces which have been polished to a mirrorlike finish, since these surfaces will photograph black with the fingerprint camera.

Black powder should be applied to white or light-colored surfaces.

Aluminum powder affords the same contrast as the gray. Gold and red bronze powders, although of a glittering appearance, will photograph dark and should consequently be used on light-colored surfaces. Dragon's blood powder is a photographically neutral powder and may be dusted on either a light or dark surface.

On clear transparent glass, either gray or black powder may be used, it being necessary only to use a contrasting black or white background when photographing.

Prints should be lifted after photographing. Both rubber and transparent tape are available for this purpose. Rubber lifting tape is procurable in black or white 4" x 9" sheets and has the adhesive surface protected with a celluloid cover. A black powder print should obviously be lifted on white tape and a gray powder print on black tape.

Gold bronze and red bronze powders should be lifted on white tape, aluminum on black. Dragon's blood may be lifted on either black or white.

After cutting a piece of tape sufficiently large to cover the entire latent print, the celluloid covering is removed and the adhesive side applied to the latent. The tape should be pressed evenly and firmly to the surface, taking care not to shift its position. It is then peeled gently from the surface and the piece of celluloid placed over the print to protect it. The operator should handle the lift in such a manner that he will leave no prints of his own on the adhesive surface. A small paper identification tag bearing the initials of the operator, date, and object from which lifted should be placed under one corner of the celluloid, or this information may be written on the back of the lift itself if it can be done in a permanent, legible manner.

If an excessive amount of powder adheres to the latent print, a more legible print may sometimes be obtained by lifting a second time (on a new piece of tape, of course).

It should be noted that a print lifted on rubber tape is in a reverse position. Consequently, in preparing a photograph of a print on such a *lift, it will be necessary to print the negative from the reverse side in order for the print to appear in its correct position for comparison. Preparation of such photographs should not be attempted by persons of inadequate knowledge and experience.*

Transparent tape with a durable adhesive surface is available in 1" to 2" widths for fingerprint work. The common variety of transparent tape is not suitable due to the deterioration (drying) of the adhesive surface. The print on a piece of transparent tape is in correct position. Transparent lifts should be affixed to a smooth, grainless, opaque background of a black or white color contrasting with the powder used. Every effort should be made to avoid air bubbles under such lifts. In no instance should a transparent lift ever be folded back on itself or stuck to another piece of such tape as a backing, since it is generally not possible to determine the correct position of such a print.

Groups of latent impressions, such as those of adjacent fingers or fingers and palms which appear to have been made simultaneously, should be lifted as units, that is, on a single piece of tape, as this may facilitate the task of making comparisons.

CHAPTER XIV

Chemical Development of Latent Impressions

Chemical treatment in the development of latent finger impressions on paper, cardboard, and newly finished or unpainted wood may involve a slightly more complicated technique than that in which powders are utilized, but the results justify the additional effort.

It is very strongly recommended that powders not be applied to articles of the above types. This recommendation is made for several reasons. First, powders cannot be removed from paper and possibly may interfere with some types of document examinations. In this connection, they are likely to prevent restoration of the specimen to a legible condition. Powders will not develop as many latent impressions as chemicals on paper or cardboard. In some cases they will obscure latent impressions subsequently developed chemically.

Neither scientific training nor complete knowledge of the chemical processes involved is necessary for one to become proficient in the use of chemical developers, two of which will be discussed more fully. These two, iodine and silver nitrate, are the most commonly used, inasmuch as they are relatively inexpensive, readily procurable, effective, and easy to apply.

All specimens which are treated should be handled with tweezers or gloves.

When iodine crystals are subjected to a slight amount of heat they vaporize rapidly, producing violet fumes. These fumes are absorbed by fatty or oily matter with which they come in contact. If the specimen treated bears latent impressions which contain oil or fat, the print is developed or made visible by the absorption of the iodine fumes and the ridges of the print appear yellowish-brown against the background.

Iodine prints are not permanent and begin to fade once the fuming is stopped. It is necessary, therefore, for the operator to have a camera ready to photograph the prints immediately.

Control of the fumes is achieved by using the crystals in an iodine gun or fuming cabinet. The iodine gun may be assembled by the individual examiner, by a druggist, or it may be purchased through a fingerprint supply house.

Material for making the iodine gun, as well as iodine crystals, may be procured from a chemical supply house or through a druggist. The gun itself consists essentially of two parts. One tube (the end of the gun through which the breath is blown) contains a drying agent such as calcium chloride, to remove moisture from the breath. Without this, the moisture from the breath and saliva would condense at the end of the gun, drip onto the specimen and cause stains which might prove indelible. The second tube contains a small amount of iodine crystals which are vaporized by the heat of the breath, augmented by the warmth of the hand cupped around the tube containing the iodine. This vapor is blown onto the specimen (fig. 420). Glass wool serves to hold the calcium chloride and iodine in place.

[Illustration: 420. Iodine fuming gun in use.]

Due to the amount of physical exertion involved, the gun is generally limited to the examination of a few small specimens. Where a large number of specimens are to be treated, the fuming cabinet, a box-shaped wooden receptacle with a glass front and top permitting the operator to control the amount of fumes in the cabinet and observe the development of the latent impressions, is used (fig. 421). The fumes are generated by placing a small alcohol burner under an evaporating dish containing the iodine crystals. This is set in a hole cut in the bottom of the cabinet. As soon as the fumes begin to appear in sufficient amounts, the burner is removed. The specimens may be hung in the cabinet by wooden clothes pins fastened to a removable stick which is supported by wooden strips affixed near the top edges of the cabinet. The top of the cabinet is removable to permit access. Diagrams for the construction of the iodine gun or fuming cabinet will be furnished on request to members of the law enforcement profession.

Many specimens bear small, greasy areas which, in addition to any latent impressions of a greasy nature, will also appear yellowish-brown

CHAPTER XIV

after exposure to iodine fumes. All these stains will eventually disappear if the specimen is placed in a current of air from a fan or vent. All latent impressions on an object will not be developed by the iodine process but only those containing fat or oil. Due to this fact and the fact that iodine evaporates from the surface, it is used prior to (it cannot be used afterward), and in conjunction with, the silver nitrate process.

[Illustration: 421. Iodine fuming cabinet in use.]

No ill effects have been noted from contact with small amounts of iodine vapor but prolonged or excessive contact will produce irritation of the skin and respiratory passages. To prevent gradual loss of the chemical through evaporation and the corrosion of surrounding metal surfaces, iodine crystals should be kept in an airtight container when not being used.

The development of latent impressions with silver nitrate is dependent on the fact that the sodium chloride (the same substance as common table salt) present in the perspiration which forms the ridges in most latent impressions reacts with the silver nitrate solution to form silver chloride. Silver chloride is white but is unstable on exposure to light and breaks down into its components, silver and chlorine. The ridges of the fingerprints developed in this manner appear reddish-brown against the background. Immersion in the silver nitrate solution will wash traces of fat and oil from the paper; consequently, it is necessary to fume the specimen for latents of such a nature prior to treatment with silver nitrate.

Once the requisite equipment is assembled, the steps in the process are these:

Dip the specimen in the solution, blot and dry it, expose to light, and photograph latents when contrast is good.

Chemically standardized solutions are not required for the successful application of this process. It has been determined through long practice that a 3-percent solution of silver nitrate is adequate for the purpose, although concentrations up to 10 percent are sometimes used. A solution of approximately 3 percent may be prepared by dissolving 4 ounces of silver nitrate in 1 gallon of distilled water. Smaller quantities of 3-percent solution are made by using the components in the same proportion. For instance, one quart of water will require 1 ounce of the crystals. For a 10-percent solution, use 13-1/3 ounces of crystals per gallon.

An alcohol solution may be preferred. This is prepared by mixing 4 ounces of silver nitrate crystals, 4 ounces of distilled water, and 1 gallon of grain alcohol, 190 proof. The alcoholic solution dries faster, and when treating paper bearing writing in ink, it is less likely to cause the ink to run. On the other hand, the alcoholic solution is much more expensive and there is some loss by evaporation while in use.

The solutions may be used several times before losing their strength and when not in use should be kept in brown bottles in cupboards to retard deterioration. If the strength of the solution is doubtful, the operator should attempt to develop test latent impressions before proceeding on evidence.

Silver nitrate crystals and distilled water in small amounts are obtainable from druggists or in large amounts from chemical supply houses. Dealers in distilled water are located in many communities.

Tap water should not be used in the preparation of the solution because it generally contains chemicals which will partially neutralize the silver nitrate.

It is suggested that the solution be placed in a glass or enamelware tray approximately 18 by 12 by 5 inches for use, a size used in photographic development. Treatment with this solution is called "silvering." The specimen is immersed in the solution so that the surfaces are completely moistened, then taken out, placed between blotters to remove the excess solution, and dried. The drying is readily accomplished with an electric hair dryer. Blotters may be dried and used several times before discarding. It is not necessary to work in a dark room. Work in an illuminated room but not in direct sunlight. Soaking the specimen in the solution does not aid development and is actually undesirable as it requires a longer drying time. The specimen should be reasonably dry before exposing to the light, otherwise the latent prints may be developed while the paper is still wet, thus necessitating drying in subdued light to prevent darkening.

Development of the latent impressions occurs rapidly when the specimen is exposed to a blue or violet light source. A 1,000-watt blue or daylight photographer's lamp, a mercury arc (most ultraviolet lamps are of this type), or carbon arc is excellent for the purpose (fig. 422). If a weaker light is used, a stronger mixture of the solution should be prepared. For instance, if a 300-watt bulb is used, the 10-percent solution would be preferable. Direct sunlight will cause the latent impressions to appear very rapidly and if several specimens are exposed at once it is not possible for a single operator to properly control the development. Sunlight coming through a window pane will serve for development. Where fingerprints containing sodium chloride (normally exuded from the sweat pores in the ridges) have been deposited, the silver chloride formed will darken against the background.

[Illustration: 422. Developing silver nitrate prints using 1,000-watt bulb reflector.]

As soon as the ridge detail of the prints is clearly visible, the paper should be removed from the light. Continued exposure will darken the paper and the contrast will be lost.

Paper so treated should be kept in darkness; that is, in a heavy envelope or drawer until ready to photograph.

Immediate photographing, as in the case of iodine prints, is not always essential, since the prints are permanent and become illegible only through eventual clouding of the background. Prompt photographing is recommended, however, as, in exceptional instances, silver-nitrate prints have become illegible in a matter of hours. Darkening ordinarily will occur slowly if the paper is preserved in absolute darkness, and silver-nitrate prints so preserved more than 10 years have been observed to be quite clear.

CHAPTER XIV

Items such as cardboard cartons, newspapers, road maps, large pieces of wrapping paper, or smooth, unpainted wood surfaces, too large for dipping, may be treated by brushing the solution over the surface with a paint brush (fig. 423). Brushing does not damage or destroy latent impressions on surfaces of this type. Cardboard boxes may be slit down the edges and flattened out to permit easy placement under the light.

[Illustration: 423. Silver nitrate solution being applied with paint brush.]

Wet paper should be handled with extreme care to prevent tearing. In treating very thin types of paper the solution is best applied with a cotton swab or brush.

Photographs, Photostats, and blueprints of any value should not be treated with silver nitrate, since the developed prints or stains cannot be removed without destroying them.

In working with silver nitrate, wear rubber gloves or handle all specimens with tweezers; avoid spilling it on clothing. It will cause dark brown stains on clothing, skin, and fingernails. Such stains are not easily removed. Areas of the skin subjected to prolonged contact are deadened, will turn black and peel.

If removal of silver nitrate prints (called "de-silvering") is desired, this may be accomplished by placing the specimen in a 2-percent solution of mercuric nitrate in a tray similar to that used for the silver nitrate.

To prepare a small amount of this solution, dissolve two-thirds of an ounce of mercuric nitrate crystals in 1 quart of distilled water and add one-third of a fluid ounce of nitric acid. Shake well. This solution, too, may be used several times before losing its strength and is not necessarily discarded after each use. It is not necessary to keep it in a dark bottle.

The specimen bearing silver nitrate prints is immersed in this solution until all traces of the prints disappear. It should then be rinsed thoroughly in water to remove all mercuric nitrate. If this is not done the paper deteriorates, becoming brittle and crumbly. A tray of distilled water may be used for rinsing or a tray of ordinary tap water changed several times during the rinsing. The specimen is then laid out flat to dry.

Wrinkles, such as are left in paper after ordinary drying, may be prevented by ironing with a moderately hot iron. An electric iron with a temperature control is desirable. If kept too hot it will scorch or wrinkle the paper somewhat. The bottom of the iron should be clean so that unremovable smudges will not be left on the paper.

No ill effects have been noted from working in the 2-percent mercuric nitrate solution with bare hands for very short periods, but it is a caustic solution and it is suggested that the specimens be handled with tweezers or that rubber gloves be worn if contact is prolonged.

CHAPTER XV

The Use of the Fingerprint Camera

If a fingerprint is visible, an effort should be made to photograph it before any attempt is made to develop it. In every case a print developed with powder should be photographed before lifting. It sometimes happens that the print does not lift properly although it may be quite clear after development.

The camera which is especially adapted to the purpose and which is easiest to handle and operate is the fingerprint camera, one type of which is shown in figure 424. This camera has several advantages in photographing fingerprints:

It photographs the prints in natural size. It contains its own light source. It has a fixed focus.

Cameras of this type are available in models operated by batteries and 110-volt current. It is believed that the battery-operated type has the greater utility, since house current may not be available at the crime scene. When not in use the batteries should be removed as they will eventually deteriorate and corrode the brass contacts in the camera.

[Illustration: 424. The fingerprint camera.]

The camera is of the box type and has three button controls which will open: (1) The metal flap covering the aperture, (2) the front portion of the frame providing access to the self-contained light bulbs, and (3) the camera in half, providing access to the batteries and the shutter as shown below in figure 425.

[Illustration: 425. Button controls permit access to bulbs, batteries, and shutter.]

A 2-1/4 x 3-1/4 film pack adapter or a 2-1/4 x 3-1/4 cut film holder holds the film in the camera. The film pack adapter will hold a pack of 12 sheets of film, and accordingly, will permit the taking of 12 pictures. The cut film holder is a unit which holds two sheets of film utilizing

CHAPTER XV

each side of the holder.

It is pointed out that the FBI uses the film pack exclusively in view of the fact that practically all latent examinations will necessitate the taking of more than two pictures. Further, the film pack is made so that it may be loaded into the adapter in the open light. Also, the films are numbered 1 through 12, which is a valuable feature in that in maintaining notes concerning the latent examinations it is a simple matter to note by the number of the negative where the latent impression was developed and photographed. Should it happen that during a latent examination all twelve of the films are not used, the film pack, with the slide in place, is taken into the darkroom and only those films which have been exposed are removed and developed. The unexposed films remain in the film pack adapter with the slide in and may be used later.

As was previously mentioned, the camera has a fixed focus; that is, the camera will take a legible picture only when the latent print is at the focal point, or exactly flush with the opening of the camera. The latent print must not be inside the open end of the camera, nor must it be beyond; otherwise, the picture will be blurred.

[Illustration: 426. When object being photographed does not cover camera opening, outside light is excluded with piece of cloth.]

Since the camera has its own light source, any leakage of outside light will cause overexposure of the film. Consequently, if the surface of the object bearing the latent print which is to be photographed is uneven or does not cover the entire front of the camera opening, it will be necessary to use some opaque material such as a focusing cloth or heavy dark material to cover the front of the camera so as to exclude all outside light (fig. 426). If a latent print on a pane of glass or an automobile window is being photographed, it will be necessary to back up the latent so that there will be no light leakage. Material showing a pattern or grain should not be used for this purpose as any such pattern will photograph in the background and possibly obliterate the ridges of the latent print.

CHAPTER XV 165

To insure an equal distribution of the light over the latents being photographed, the impressions should be centered in the opening of the camera. This is accomplished by opening the angular front section of the camera after the metal plate covering the front has been opened, and setting the aperture over the latent impressions so that they will be as near the center as possible. Then, holding the camera firmly in place, it is carefully closed (fig. 427).

During exposure the camera must be held perfectly still. Any movement of the camera or object will result in a fuzzy or double image.

In photographing a small, movable object such as a bottle or tumbler, the camera should not be placed on its end and an attempt made to balance the object across the opening. Instead, the camera should be placed on its side and the bottle or tumbler built up to the opening so that there is no necessity for holding the object (fig. 428). There will be, of course, instances where the camera will have to be held, such as to the side of a wall, cabinet or automobile. Here an extreme effort should be made to avoid moving the camera or permitting it to slip during exposure.

[Illustration: 427. Centering of latent in aperture insures equal distribution of light over print.]

Anticipating the possible use of the photographic negatives in a court proceeding, it becomes of paramount importance to be able to identify them. This is done by using what is called an identification tag. The tag consists of a small piece of paper bearing the date, initials of the examiner, and possibly a case number, and it should be hand-written. The tag is placed near the latent prints being photographed so that it will appear in the picture. It should be borne in mind that concentration should be on the latent impressions, which must be centered, and the identification tag should be to one side and not covering any of the latent prints. Another method of identification, if the surface permits, is to write the above-mentioned data on the surface of the object bearing the latents so that the information set out will also be a part of the picture. Too much emphasis cannot be placed on the importance of the identification tag. The lack of such data, by discretion of the court, may exclude the latents as evidence, in the absence of the original specimen bearing the latents.

[Illustration: 428. In photographing objects with curved or irregular surfaces, camera should be laid flat and latent-bearing surface built up to opening.]

The following are suggested exposure times for Tri-X film (available in 2-1/4" x 3-1/4" film packs) with battery-operated cameras having lenses without diaphragms:

Black powder print on white or light background--snapshot Gray or white powder print on black or dark background--1 second

These same exposure times can be used on some cameras having lenses with diaphragms, provided the lens opening is set at f6.3.

The above exposure times are for cameras with batteries in average condition. If batteries become weak the exposures may be increased slightly.

In making snapshots the shutter mechanism should be manipulated as rapidly as possible since slow motion will appreciably lengthen the exposure. In making time exposures the camera shutter must be held open for the desired time. Personnel with photographic experience may desire to use cut film with the fingerprint camera. A few tests will determine the optimum exposure times for any particular type of film.

Briefly, the procedure for taking photographs of latents is as follows: The film pack is placed into the film pack adapter with the safety paper side of the film pack to the slide side of the adapter, care being taken to see that all of the paper tabs are outside of the adapter. The adapter is placed on the camera in its proper position by opening a slide clamp attached to the camera, fitting the side of the adapter into the slot away from the clamp and pushing it down flat into the opening. Don't try to slide the adapter into the opening from the top. The adapter is locked in position by closing the clamp. Next the slide is removed and the tab marked "Safety Cover" pulled out as far as it will come and torn off. The camera is placed in position and the first exposure made; then the

CHAPTER XV

tab marked "1" is pulled and the next film is in position for exposure. This procedure is followed with each succeeding film until all have been exposed.

When the last tab has been pulled out, the pack can be removed from the adapter in daylight. If all of the films have not been exposed the slide is replaced into the adapter and the film pack removed from it in a darkroom, as previously stated.

As a matter of regular policy, it is recommended that more than one exposure be made of each latent, varying the normal exposure time to insure satisfactory results, especially when the contrast is not a good black on white or gray on black.

Before starting to photograph, note the following:

- Check shutter action.

- Check bulbs, batteries, and lights.

- Center latents in opening of camera.

- Latent being photographed must be flush with opening of camera.

- Outside light must be excluded.

- Include identification tag in photograph.

- Remove slide and pull safety tab of film pack before making exposure.

- Hold camera still while making exposure.

- Pull correct number tab after each exposure (be careful not to pull more than one).

- Do not use grained or uneven material as a backing when photographing latents on transparent glass.

CHAPTER XV

- Mirrors, polished chrome, and nickel plate will photograph black in the fingerprint camera.

The foregoing has dealt with the standard use of the fingerprint camera when the direct light afforded by the camera gives suitable results. There will be cases, however, where the results from the use of the direct light may not be adequate. Such cases may involve molded or embedded prints, such as prints in putty, wax, soap, etc. Should direct light give poor results, side lighting may prove effective. This can be done by loosening two of the bulbs on one side so that they will not light. The light given by the other two bulbs is directed so as to pass at right angles, as much as possible, across the ridges of the embedded latent print. Adjustment of the exposure time must be made when this is done.

CHAPTER XVI

Preparation of Fingerprint Charts for Court Testimony

In testifying to fingerprint identification, the expert often prepares charts to visually aid the court and jury in understanding the nature of his testimony. Many times it is undoubtedly difficult for the layman to perceive, from a vocal explanation alone, the full import of an expert's testimony, due to its technical nature; consequently, some graphic representation of the facts presented is amply justified and rewarded. The preparation of the charts is ultimately the sole responsibility of the expert using them. As a matter of interest to law enforcement personnel engaged in fingerprint work, a brief explanation of the preparation of such charts follows, along with suggestions and remarks based on long experience in these matters.

To do the work conveniently, it will be necessary to have available, in addition to the ordinary photographic developing and printing materials, a projection enlarger which will enlarge preferably to at least ten diameters. In the projection method of enlargement, the image is printed directly from the original negative, and the preparation of an enlarged negative is unnecessary.

Aside from the photographic equipment, the needed materials are: a roll of scotch photographic tape 1 inch wide to outline the areas of the fingerprints on the negatives to be used; some stiff cardboard approximately 1/32 inch thick on which to mount the prepared charts; a tube of rubber cement; and a bottle of translucent ink, other than black or white.

A light-box on which to view the negatives while blocking, and a lettering set to draw the lines and numbers uniformly on the charts, while not absolutely essential, are helpful conveniences. A light-box is basically a frosted pane of glass with a light beneath it to produce soft, even, non-glaring illumination. If no light-box is available, a clear window may be utilized in "blocking" the negatives.

If the expert finds it necessary to have an outside source prepare his photographs, he should retain personal custody of the evidence during the operation.

The original latent print and inked print with which it is identical should be photographed actual size. This procedure eliminates guesswork in enlarging both to the same degree. Whatever areas of the two prints are deemed requisite to illustrate the method of identification are then outlined (blocked) on the negatives with the masking tape, so that only those areas will show in the subsequent enlargements. Generally, if the legible area of the latent print is small, it is well to show the complete print. If the area is large, however, as in a palm print, an area which will not make the chart too bulky or unwieldy may be selected.

In blocking, the negative is affixed to the window pane or light-box by means of strips of photographic tape across the corners, with the side to be blocked up. This prevents constant shifting of the negative while it is being prepared. The latent print should be blocked first. Corners of the blocked areas should be square. Care should be exercised to have as nearly as possible the same ridge formations shown and the ridge formations in the same upright or horizontal positions. This may be facilitated by fixing a negative, bearing ruled squares, between the negative being blocked and the glass to which it is attached.

If the latent print was developed or photographed as a light print on a dark background, a reverse-color negative should be prepared and blocked in order that both prints may appear as black ridges on light backgrounds. This is done by placing the original negative adjacent to a new sheet of film and exposing it. The resultant negative contains the same image as the original except that the color of the image has been reversed.

If the negative is a photograph of an opaque lift, the print appears in reverse position; that is, as a mirror image, and the negative will accordingly have to be blocked from the dull or emulsion side in order for it to appear in a position comparable to that of the inked print.

Failure to present the prints in question in the same color and position may possibly confuse the observer and nullify the purpose for which the chart is made.

CHAPTER XVI

The degree of enlargement is not important in itself, so long as the ridges of the latent print are readily distinguishable by the eye. Ten diameters have been found adequate, although any enlargement from 5 to 30 will serve. It should be remembered, however, that small enlargements are difficult to see a few feet away and that large ones lose some of the contrast between ridges and background. A white border of at least 1-1/2 inches or a width equal to about one-third the enlarged area should be left for charting purposes.

Any chart prepared must be technically correct; that is, the corresponding ridge characteristics in the two prints must be similarly numbered and indicated.

Several ways of pointing out the similar ridge formations have been observed, but the one which appears soundest is also simplest and consists of merely marking the characteristics with lines and numbers.

All of the ridge characteristics in the prints need not be charted. Twelve characteristics are ample to illustrate an identification, but it is neither claimed nor implied that this number is required.

All fingerprint identifications are made by observing that two impressions have ridge characteristics of similar shapes which occupy the same relative positions in the patterns.

Methods involving superimposition of the prints are not recommended because such a procedure is possible only in a very few instances, due to the distortion of ridges in most prints through pressure and twisting. Such a procedure is not necessarily a test of identity.

Likewise, presenting charts with the shapes of the characteristics drawn in the margin is not recommended. Individual ridge characteristics may vary slightly in actual shape or physical position due to twisting, pressure, incomplete inking, condition of latent print when developed, powder adhering to background, etc. Identifications are based on a number of characteristics viewed in a unit relationship and not on the microscopic appearances of single characteristics.

CHAPTER XVI

Since the enlarged photographs appear in black and white, an ink other than black or white should be used to line the chart. Such an ink should be preferably translucent so that it will be possible to see the ridges which it traverses. A translucent carmine drawing ink serves well. In placing the lines on the chart, they should be arranged so that they do not cross or touch.

The chart will present a clearer, neater, and more pleasing appearance if it is numbered clockwise and the numbers are evenly spaced (fig. 429). It is not necessary, however, to place the numbers evenly around the photograph.

[Illustration: 429. Chart illustrating method of fingerprint identification.]

Ordinarily, the numbers are placed on three sides and the type of print (latent or ink) noted at the bottom. In any case, the manner of numbering should be subservient to an explanation of the characteristics in an orderly sequence; and, if the situation warrants it, all of the points may be illustrated on a single side of the photograph.

A single line should be drawn from each characteristic to a numbered point on the margin. Care should be taken to draw the line exactly to the characteristic point, not short of it, beyond it, or obscuring it. Erasures should be avoided. If the ink runs or blots, it is sometimes possible to remove it with a cloth dampened in denatured alcohol, without damaging the photograph.

If the enlargement is great, that is, 25 or 30 diameters, it might be well to draw a small circle around each characteristic and then draw the line from the circle to the number, since the ridge will be much thicker than the illustrating line. All lines and numbers should be checked for absolute accuracy. The expert should also study the enlargements for apparent discrepancies in the prints, which he might be called upon to explain.

The charted enlargements are readily mounted on stiff cardboard with rubber cement, which may be purchased in small tubes. After cementing the photograph to the cardboard, it should be placed under a heavy flat object which will cover the entire surface until dry to prevent warping and wrinkling. After drying, trim the two enlargements to the same square size with heavy scissors, a pen knife or scalpel, and fasten them together, book-fashion, with strips of the photographic tape used in blocking the negatives. Of course, if charts are large, 20 to 36 inches square, mounting is unnecessary and they will have to be supported in the courtroom with thumbtacks or metal rings.

Some courts do not permit numbering or lining of the photographs and the enlargements alone in these cases will have to suffice. If there is some question about admissibility of the charted enlargements, it is well to prepare an extra uncharted set.

CHAPTER XVII

Unidentified Latent Fingerprint File

From time to time the FBI is requested to conduct surveys and participate in conferences and in police schools on the problem of fingerprint identification.

As a result of its observations in the course of these activities it has been found that many identification bureaus are not fully aware of the importance which latent finger impressions can have in connection with the ordinary handling of arrest fingerprint cards.

Many bureaus and departments spend considerable time in developing latent impressions in a particular case. If no immediate results are forthcoming, the latent impressions are filed for future reference.

Single fingerprint files have been maintained with success by some departments. Many others do not attempt to keep a file because of either limited personnel or lack of funds. In many departments, however, where such a file is maintained, too often latent impressions are simply filed with no regard to possible future use. Actually, these impressions should be treated as evidence directly connecting the subject with the crime.

Active consideration should be given to the latent impressions until they are identified or the case has been successfully prosecuted. It is definitely felt that the following suggested procedure might have some decided advantages.

It is suggested that in all cases where latent impressions are developed at the crime scene, or on an object used in connection with the commission of a crime, the impressions be properly photographed and lifted. The evidence, where possible and practicable, should be properly packed, labeled, and stored for future use in court (fig. 430).

Use care in wrapping the evidence to see that the latent impressions on the objects are not destroyed. If the specimens are later used in court, the impressions should still be clearly visible. In the same manner, all evidence not of a bulky nature, such as photographic negatives, photographs, and lifts of latent impressions, should be similarly preserved for future court use. It is to be emphasized that all material in one case should bear a case number. All specimens not of a bulky nature can be placed in an envelope and filed by this case number (fig. 431).

The above procedure is the usual one followed by the majority of identification bureaus in handling latent impressions. In order, however, to keep the latents in an active state, the photographs of all the latent impressions found in a particular case should be cut up and pasted on a 3 by 5 card bearing the case number and title of the case (fig. 432).

[Illustration: 430. Evidence labeled and latents protected for storing for future court use.]

[Illustration: 431. Latent material in a case should be filed under a single case number.]

If numerous latents are developed, several cards should be used, all having the same number and title. These cards are then filed by case number in a regular filing cabinet. Before this step is taken, every effort should be made to secure and compare the fingerprints of individuals who may legitimately have placed their prints on the objects which were examined. In addition, as part of the case report bearing the same case number as the latent impressions, there should be a notation pointing out that latent impressions were developed in the case and that they are on file.

[Illustration: 432. For ready current comparisons latents in a case are placed on a 3 x 5 card bearing case title and number.]

Case #2345

Unknown Subjects Jones' Drug Co. B&E 3-15-47

Fingerprint comparisons in this unidentified file can be made on the basis of fingerprints taken from day to day of individuals fingerprinted *for criminal identification purposes. A routine may be set up whereby the fingerprints of individuals arrested each day will be compared the following day with the latent fingerprints filed in the unidentified latent file. It is most important that this procedure be rigidly followed from day to day. It is to be borne in mind that the comparisons are made whether the particular person is or is not a suspect in a certain case. Special attention should be paid to fingerprints of individuals charged with burglary, breaking and entering, armed robbery, and other similar crimes.*

Should an identification be made of some latent prints, and others in the same case remain unidentified, the 3 by 5 card should remain in file until the case is fully closed, inasmuch as more than one person may be involved in the crime. Of course, if all the latents are identified, then the 3 by 5 card is removed and placed with the negatives, lifts, etc.

It may be deemed advisable to remove these latents from the file in instances where the statute of limitations covering the crime has run.

If the above procedure is rigidly followed, identification in many instances will result--more than would be effected if the department maintained only a single fingerprint file in which the latent prints were merely filed away. Very often such a latent fingerprint file is a source of information when all logical investigative leads have been exhausted.

This résumé of latent impressions has been prepared by the Federal Bureau of Investigation in the belief that it may be of possible interest to law enforcement officers desiring to avail themselves of latent identification evidence in connection with their investigative activities. It should be borne in mind that the comments and expressions set out in this book are not intended to convey the thought that the Federal Bureau of Investigation believes the points emphasized are the only ones of moment, or that other methods of developing latent impressions are not equally acceptable. The Federal Bureau of Investigation will be glad to answer any questions on the foregoing which may occur to any law enforcement officer who reads this material._

End of the Project Gutenberg EBook of The Science of Fingerprints, by Federal Bureau of Investigation John Edgar Hoover

*** END OF THIS PROJECT GUTENBERG EBOOK THE SCIENCE OF FINGERPRINTS ***

***** This file should be named 19022-8.txt or 19022-8.zip ***** This and all associated files of various formats will be found in:
http://www.gutenberg.org/1/9/0/2/19022/

Produced by Jason Isbell, Linda Cantoni, and the Online Distributed Proofreading Team at http://www.pgdp.net

Updated editions will replace the previous one--the old editions will be renamed.

Creating the works from public domain print editions means that no one owns a United States copyright in these works, so the Foundation (and you!) can copy and distribute it in the United States without permission and without paying copyright royalties. Special rules, set forth in the General Terms of Use part of this license, apply to copying and distributing Project Gutenberg-tm electronic works to protect the PROJECT GUTENBERG-tm concept and trademark. Project Gutenberg is a registered trademark, and may not be used if you charge for the eBooks, unless you receive specific permission. If you do not charge anything for copies of this eBook, complying with the rules is very easy. You may use this eBook for nearly any purpose such as creation of derivative works, reports, performances and research. They may be modified and printed and given away--you may do practically ANYTHING with public domain eBooks. Redistribution is subject to the trademark license, especially commercial redistribution.

*** START: FULL LICENSE ***

THE FULL PROJECT GUTENBERG LICENSE PLEASE READ THIS BEFORE YOU DISTRIBUTE OR USE THIS WORK

To protect the Project Gutenberg-tm mission of promoting the free distribution of electronic works, by using or distributing this work (or any other work associated in any way with the phrase "Project Gutenberg"), you agree to comply with all the terms of the Full Project Gutenberg-tm License (available with this file or online at http://gutenberg.org/license).

Section 1. General Terms of Use and Redistributing Project Gutenberg-tm electronic works

1.A. By reading or using any part of this Project Gutenberg-tm electronic work, you indicate that you have read, understand, agree to and accept all the terms of this license and intellectual property (trademark/copyright) agreement. If you do not agree to abide by all the terms of this agreement, you must cease using and return or destroy all copies of Project Gutenberg-tm electronic works in your possession. If you paid a fee for obtaining a copy of or access to a Project Gutenberg-tm electronic work and you do not agree to be bound by the terms of this agreement, you may obtain a refund from the person or entity to whom you paid the fee as set forth in paragraph 1.E.8.

1.B. "Project Gutenberg" is a registered trademark. It may only be used on or associated in any way with an electronic work by people who agree to be bound by the terms of this agreement. There are a few things that you can do with most Project Gutenberg-tm electronic works even without complying with the full terms of this agreement. See paragraph 1.C below. There are a lot of things you can do with Project Gutenberg-tm electronic works if you follow the terms of this agreement and help preserve free future access to Project Gutenberg-tm electronic works. See paragraph 1.E below.

1.C. The Project Gutenberg Literary Archive Foundation ("the Foundation" or PGLAF), owns a compilation copyright in the collection of Project Gutenberg-tm electronic works. Nearly all the individual works in the collection are in the public domain in the United States. If an individual work is in the public domain in the United States and you are located in the United States, we do not claim a right to prevent you from copying, distributing, performing, displaying or creating derivative works based on the work as long as all references to Project Gutenberg are removed. Of course, we hope that you will support the

CHAPTER XVII

Project Gutenberg-tm mission of promoting free access to electronic works by freely sharing Project Gutenberg-tm works in compliance with the terms of this agreement for keeping the Project Gutenberg-tm name associated with the work. You can easily comply with the terms of this agreement by keeping this work in the same format with its attached full Project Gutenberg-tm License when you share it without charge with others.

1.D. The copyright laws of the place where you are located also govern what you can do with this work. Copyright laws in most countries are in a constant state of change. If you are outside the United States, check the laws of your country in addition to the terms of this agreement before downloading, copying, displaying, performing, distributing or creating derivative works based on this work or any other Project Gutenberg-tm work. The Foundation makes no representations concerning the copyright status of any work in any country outside the United States.

1.E. Unless you have removed all references to Project Gutenberg:

1.E.1. The following sentence, with active links to, or other immediate access to, the full Project Gutenberg-tm License must appear prominently whenever any copy of a Project Gutenberg-tm work (any work on which the phrase "Project Gutenberg" appears, or with which the phrase "Project Gutenberg" is associated) is accessed, displayed, performed, viewed, copied or distributed:

This eBook is for the use of anyone anywhere at no cost and with almost no restrictions whatsoever. You may copy it, give it away or re-use it under the terms of the Project Gutenberg License included with this eBook or online at www.gutenberg.org

1.E.2. If an individual Project Gutenberg-tm electronic work is derived from the public domain (does not contain a notice indicating that it is posted with permission of the copyright holder), the work can be copied and distributed to anyone in the United States without paying any fees or charges. If you are redistributing or providing access to a work with the phrase "Project Gutenberg" associated with or appearing on the work, you must comply either with the requirements of paragraphs 1.E.1 through 1.E.7 or obtain permission for the use of the work and the Project Gutenberg-tm trademark as set forth in paragraphs 1.E.8 or 1.E.9.

1.E.3. If an individual Project Gutenberg-tm electronic work is posted with the permission of the copyright holder, your use and distribution must comply with both paragraphs 1.E.1 through 1.E.7 and any additional terms imposed by the copyright holder. Additional terms will be linked to the Project Gutenberg-tm License for all works posted with the permission of the copyright holder found at the beginning of this work.

1.E.4. Do not unlink or detach or remove the full Project Gutenberg-tm License terms from this work, or any files containing a part of this work or any other work associated with Project Gutenberg-tm.

1.E.5. Do not copy, display, perform, distribute or redistribute this electronic work, or any part of this electronic work, without prominently displaying the sentence set forth in paragraph 1.E.1 with active links or immediate access to the full terms of the Project Gutenberg-tm License.

1.E.6. You may convert to and distribute this work in any binary, compressed, marked up, nonproprietary or proprietary form, including any word processing or hypertext form. However, if you provide access to or distribute copies of a Project Gutenberg-tm work in a format other than "Plain Vanilla ASCII" or other format used in the official version posted on the official Project Gutenberg-tm web site (www.gutenberg.org), you must, at no additional cost, fee or expense to the user, provide a copy, a means of exporting a copy, or a means of obtaining a copy upon request, of the work in its original "Plain Vanilla ASCII" or other form. Any alternate format must include the full Project Gutenberg-tm License as specified in paragraph 1.E.1.

1.E.7. Do not charge a fee for access to, viewing, displaying, performing, copying or distributing any Project Gutenberg-tm works unless you comply with paragraph 1.E.8 or 1.E.9.

CHAPTER XVII

1.E.8. You may charge a reasonable fee for copies of or providing access to or distributing Project Gutenberg-tm electronic works provided that

- You pay a royalty fee of 20% of the gross profits you derive from the use of Project Gutenberg-tm works calculated using the method you already use to calculate your applicable taxes. The fee is owed to the owner of the Project Gutenberg-tm trademark, but he has agreed to donate royalties under this paragraph to the Project Gutenberg Literary Archive Foundation. Royalty payments must be paid within 60 days following each date on which you prepare (or are legally required to prepare) your periodic tax returns. Royalty payments should be clearly marked as such and sent to the Project Gutenberg Literary Archive Foundation at the address specified in Section 4, "Information about donations to the Project Gutenberg Literary Archive Foundation."

- You provide a full refund of any money paid by a user who notifies you in writing (or by e-mail) within 30 days of receipt that s/he does not agree to the terms of the full Project Gutenberg-tm License. You must require such a user to return or destroy all copies of the works possessed in a physical medium and discontinue all use of and all access to other copies of Project Gutenberg-tm works.

- You provide, in accordance with paragraph 1.F.3, a full refund of any money paid for a work or a replacement copy, if a defect in the electronic work is discovered and reported to you within 90 days of receipt of the work.

- You comply with all other terms of this agreement for free distribution of Project Gutenberg-tm works.

1.E.9. If you wish to charge a fee or distribute a Project Gutenberg-tm electronic work or group of works on different terms than are set forth in this agreement, you must obtain permission in writing from both the Project Gutenberg Literary Archive Foundation and Michael Hart, the owner of the Project Gutenberg-tm trademark. Contact the Foundation as set forth in Section 3 below.

1.F.

1.F.1. Project Gutenberg volunteers and employees expend considerable effort to identify, do copyright research on, transcribe and proofread public domain works in creating the Project Gutenberg-tm collection. Despite these efforts, Project Gutenberg-tm electronic works, and the medium on which they may be stored, may contain "Defects," such as, but not limited to, incomplete, inaccurate or corrupt data, transcription errors, a copyright or other intellectual property infringement, a defective or damaged disk or other medium, a computer virus, or computer codes that damage or cannot be read by your equipment.

1.F.2. LIMITED WARRANTY, DISCLAIMER OF DAMAGES - Except for the "Right of Replacement or Refund" described in paragraph 1.F.3, the Project Gutenberg Literary Archive Foundation, the owner of the Project Gutenberg-tm trademark, and any other party distributing a Project Gutenberg-tm electronic work under this agreement, disclaim all liability to you for damages, costs and expenses, including legal fees. YOU AGREE THAT YOU HAVE NO REMEDIES FOR NEGLIGENCE, STRICT LIABILITY, BREACH OF WARRANTY OR BREACH OF CONTRACT EXCEPT THOSE PROVIDED IN PARAGRAPH F3. YOU AGREE THAT THE FOUNDATION, THE TRADEMARK OWNER, AND ANY DISTRIBUTOR UNDER THIS AGREEMENT WILL NOT BE LIABLE TO YOU FOR ACTUAL, DIRECT, INDIRECT, CONSEQUENTIAL, PUNITIVE OR INCIDENTAL DAMAGES EVEN IF YOU GIVE NOTICE OF THE POSSIBILITY OF SUCH DAMAGE.

1.F.3. LIMITED RIGHT OF REPLACEMENT OR REFUND - If you discover a defect in this electronic work within 90 days of receiving it, you can receive a refund of the money (if any) you paid for it by sending a written explanation to the person you received the work from. If you received the work on a physical medium, you must return the medium with your written explanation. The person or entity that provided you with the defective work may elect to provide a replacement copy in lieu of a refund. If you received the work electronically, the person or entity providing it to you may choose to give you a second opportunity to receive the work electronically in lieu of a refund. If the second copy is also defective, you may demand a refund in writing without further opportunities to fix the problem.

CHAPTER XVII

1.F.4. Except for the limited right of replacement or refund set forth in paragraph 1.F.3, this work is provided to you 'AS-IS' WITH NO OTHER WARRANTIES OF ANY KIND, EXPRESS OR IMPLIED, INCLUDING BUT NOT LIMITED TO WARRANTIES OF MERCHANTIBILITY OR FITNESS FOR ANY PURPOSE.

1.F.5. Some states do not allow disclaimers of certain implied warranties or the exclusion or limitation of certain types of damages. If any disclaimer or limitation set forth in this agreement violates the law of the state applicable to this agreement, the agreement shall be interpreted to make the maximum disclaimer or limitation permitted by the applicable state law. The invalidity or unenforceability of any provision of this agreement shall not void the remaining provisions.

1.F.6. **INDEMNITY**

- You agree to indemnify and hold the Foundation, the trademark owner, any agent or employee of the Foundation, anyone providing copies of Project Gutenberg-tm electronic works in accordance with this agreement, and any volunteers associated with the production, promotion and distribution of Project Gutenberg-tm electronic works, harmless from all liability, costs and expenses, including legal fees, that arise directly or indirectly from any of the following which you do or cause to occur: (a) distribution of this or any Project Gutenberg-tm work, (b) alteration, modification, or additions or deletions to any Project Gutenberg-tm work, and (c) any Defect you cause.

Section 2. Information about the Mission of Project Gutenberg-tm

Project Gutenberg-tm is synonymous with the free distribution of electronic works in formats readable by the widest variety of computers including obsolete, old, middle-aged and new computers. It exists because of the efforts of hundreds of volunteers and donations from people in all walks of life.

Volunteers and financial support to provide volunteers with the assistance they need, is critical to reaching Project Gutenberg-tm's goals and ensuring that the Project Gutenberg-tm collection will remain freely available for generations to come. In 2001, the Project

CHAPTER XVII

Gutenberg Literary Archive Foundation was created to provide a secure and permanent future for Project Gutenberg-tm and future generations. To learn more about the Project Gutenberg Literary Archive Foundation and how your efforts and donations can help, see Sections 3 and 4 and the Foundation web page at http://www.pglaf.org.

Section 3. Information about the Project Gutenberg Literary Archive Foundation

The Project Gutenberg Literary Archive Foundation is a non profit 501(c)(3) educational corporation organized under the laws of the state of Mississippi and granted tax exempt status by the Internal Revenue Service. The Foundation's EIN or federal tax identification number is 64-6221541. Its 501(c)(3) letter is posted at http://pglaf.org/fundraising. Contributions to the Project Gutenberg Literary Archive Foundation are tax deductible to the full extent permitted by U.S. federal laws and your state's laws.

The Foundation's principal office is located at 4557 Melan Dr. S. Fairbanks, AK, 99712., but its volunteers and employees are scattered throughout numerous locations. Its business office is located at 809 North 1500 West, Salt Lake City, UT 84116, (801) 596-1887, email business@pglaf.org. Email contact links and up to date contact information can be found at the Foundation's web site and official page at http://pglaf.org

For additional contact information: Dr. Gregory B. Newby Chief Executive and Director gbnewby@pglaf.org

Section 4. Information about Donations to the Project Gutenberg Literary Archive Foundation

Project Gutenberg-tm depends upon and cannot survive without wide spread public support and donations to carry out its mission of increasing the number of public domain and licensed works that can be freely distributed in machine readable form accessible by the widest array of equipment including outdated equipment. Many small donations ($1 to $5,000) are particularly important to maintaining tax

CHAPTER XVII

exempt status with the IRS.

The Foundation is committed to complying with the laws regulating charities and charitable donations in all 50 states of the United States. Compliance requirements are not uniform and it takes a considerable effort, much paperwork and many fees to meet and keep up with these requirements. We do not solicit donations in locations where we have not received written confirmation of compliance. To SEND DONATIONS or determine the status of compliance for any particular state visit http://pglaf.org

While we cannot and do not solicit contributions from states where we have not met the solicitation requirements, we know of no prohibition against accepting unsolicited donations from donors in such states who approach us with offers to donate.

International donations are gratefully accepted, but we cannot make any statements concerning tax treatment of donations received from outside the United States. U.S. laws alone swamp our small staff.

Please check the Project Gutenberg Web pages for current donation methods and addresses. Donations are accepted in a number of other ways including checks, online payments and credit card donations. To donate, please visit: http://pglaf.org/donate

Section 5. General Information About Project Gutenberg-tm electronic works.

Professor Michael S. Hart is the originator of the Project Gutenberg-tm concept of a library of electronic works that could be freely shared with anyone. For thirty years, he produced and distributed Project Gutenberg-tm eBooks with only a loose network of volunteer support.

Project Gutenberg-tm eBooks are often created from several printed editions, all of which are confirmed as Public Domain in the U.S. unless a copyright notice is included. Thus, we do not necessarily keep eBooks in compliance with any particular paper edition.

Most people start at our Web site which has the main PG search facility:

http://www.gutenberg.org

This Web site includes information about Project Gutenberg-tm, including how to make donations to the Project Gutenberg Literary Archive Foundation, how to help produce our new eBooks, and how to subscribe to our email newsletter to hear about new eBooks.

The Science of Fingerprints, by

A free ebook from http://manybooks.net/

www.ingramcontent.com/pod-product-compliance
Lightning Source LLC
Chambersburg PA
CBHW050058230526
45470CB00004B/1576